Real-Time Digital Signal Processing

Real-Time Digital Signal Processing
Based on the TMS320C6000

by Nasser Kehtarnavaz
University of Texas at Dallas

With laboratory contributions by Namjin Kim

ELSEVIER

AMSTERDAM • BOSTON • HEIDELBERG • LONDON
NEW YORK • OXFORD • PARIS • SAN DIEGO
SAN FRANCISCO • SINGAPORE • SYDNEY • TOKYO

Newnes is an imprint of Elsevier

Newnes

Newnes is an imprint of Elsevier
200 Wheeler Road, Burlington, MA 01803, USA
Linacre House, Jordan Hill, Oxford OX2 8DP, UK

 Recognizing the importance of preserving what has been written, Elsevier prints its books on
acid-free paper whenever possible.

Library of Congress Cataloging-in-Publication Data

Kehtarnavaz, Nasser.
    Real-time digital signal processing based on the TMS320C6000 / by Nasser Kehtarnavaz
; with laboratory contributions by Namjin Kim.
        p. cm.
    ISBN-13: 978-0-7506-7830-8   ISBN-10: 0-7506-7830-5
        1. Signal processing—Digital techniques. 2. Texas Instruments TMS320 series
microprocessors. I. Title

    TK5102.9.K4524 2004
    621.382'2—dc22
    ISBN-13: 978-0-7506-7830-8   2004050443
    ISBN-10: 0-7506-7830-5
British Library Cataloguing-in-Publication Data
A catalogue record for this book is available from the British Library.

For information on all Newnes publications
visit our website at www.newnespress.com

        07 08 09 10 9 8 7 6 5 4 3 2

Printed in the United States of America.

Contents

Preface

The TMS320C6000 DSP processor family has been introduced by Texas Instruments to meet high performance demands in signal processing applications. The objective of this book is to provide the know-how for the implementation and optimization of computationally intensive signal processing algorithms on the family of TMS320C6x DSP processors. In the previous version of the book named *DSP System Design: Using the TMS320C6000*, the lab exercises were based on the C6x EVM board characteristics and software. In this version named *Real-Time Digital Signal Processing Based on the TMS320C6000*, the lab exercises are redone on the C6x DSK board considering that DSK provides a more cost-effective learning platform. The migration from EVM to DSK was not a straightforward task as there were many issues that needed to be resolved, such as differences in memory maps, peripherals, host programming using the host-port interface, and issues related to the upgrading of Code Composer Studio.

The book is written so that it can be used as the textbook for real-time DSP laboratory courses offered at many schools. The material presented is primarily written for those who are already familiar with DSP concepts and are interested in designing DSP systems based on the TI C6x DSP products. Note that a great deal of the information in this book appears in the TI manuals on the C6000 DSP family. However, this information has been restructured, modified, and condensed to be used for teaching a DSP laboratory course in a semester period. It is recommended that these manuals are used in conjunction with this book to fully make use of the materials presented.

Eight lab exercises together with four project examples are discussed and included on the accompanying CD-ROM to take the reader through the entire process of C6x code writing. As a result, the book can be used as a self-study guide for implementing algorithms on the C6x DSPs. The chapters are organized to create a close correlation between the topics and lab exercises if they are used as lecture materials for a DSP lab course. Knowledge of the C programming language is required for understanding and performing the lab exercises.

Acknowledgments

I would like to express my gratitude to Texas Instruments for their permission to use the materials in their C6x manuals. All the figures marked by † in their captions are redrawn or modified at Courtesy of Texas Instruments. I extend my appreciation to Gene Frantz, TI Principal Fellow, who brought to my attention the need for writing this book. I am deeply indebted to the graduate student Namjin Kim who provided invaluable help for DSK related code modifications. Finally, I wish to thank Charles Glaser, Senior Acquisition Editor at Elsevier; Kelly Johnson, Production Editor at Elsevier; and Cathy Wicks, University Program Manager at Texas Instruments, for their support of this book.

What's on the CD-ROM?

Included on the accompanying CD-ROM:
- The lab files corresponding to the following DSP platforms:
    - o DSK6x11   : DSK 6711/6211 using AD535 on-board codec
    - o DSK6x11_ADC   : DSK 6711/6211 using PCM3003 audio daughter card
    - o DSK6416   : DSK 6416
    - o DSK6713   : DSK 6713
    - o EVM6x01   : EVM 6701/6201
    - o SIM6xxx   : Simulator

- In each platform folder, subfolders for eight labs and four projects are contained as follows:
    - o Lab01: Source files for getting familiar with Code Composer Studio.
    - o Lab02: Source files for audio sampling.
    - o Lab03: Source files for Q-format and scaling.
    - o Lab04: Source files for FIR filter.
    - o Lab05: Source files for adaptive filter.
    - o Lab06: Source files for frame processing and DMA operation.
    - o Lab07: Source files for real-time analysis.
    - o Lab08: Source files for real-time synchronization and communication.
    - o Proj01: Source files for sine wave generation.
    - o Proj02: Source files for second-order cascade IIR filter.
    - o Proj03: Source files for filter bank.
    - o Proj04: Source files for PN sequence generation.

- All the subfolders for labs and projects need to be copied into the folder "C:\ti\myprojects\" where the CCS is installed.

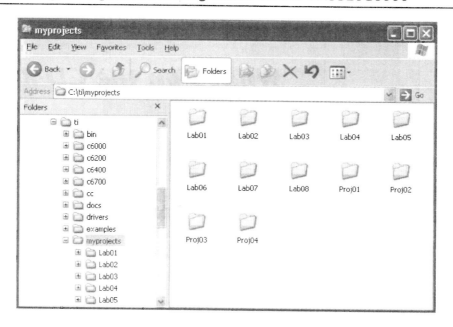

CHAPTER 1

Introduction

In general, sensors generate analog signals in response to various physical phenomena that occur in an analog manner (i.e., in continuous time and amplitude). Processing of signals can be done either in analog or digital domain. To do the processing of an analog signal in digital domain, it is required that a digital signal is formed by sampling and quantizing (digitizing) the analog signal. Hence, in contrast to an analog signal, a digital signal is discrete in both time and amplitude. The digitization process is achieved via an analog-to-digital (A/D) converter.

Digital signal processing (DSP) involves the manipulation of digital signals in order to extract useful information from them. Although an increasing amount of signal processing is being done in digital domain, there remains the need for interfacing to the analog world in which we live. Analog-to-digital (A/D) and digital-to-analog (D/A) data converters are the devices that make this interfacing possible. Figure 1-1 illustrates the main components of a DSP system, consisting of A/D, DSP, and D/A devices.

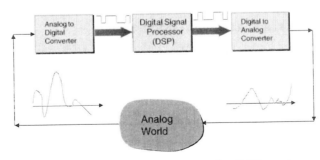

Figure 1-1: Main components of a DSP system.

There are many reasons why one would want to process an analog signal in a digital fashion by converting it into a digital signal. The main reason is that digital processing allows programmability. The same DSP hardware can be used for many different applications by simply changing the code residing in memory. Another reason is that digital circuits provide a more stable and tolerant output than analog circuits—for instance, when subjected to temperature changes. In addition, the advantage of operating in digital domain may be intrinsic. For example, a linear phase filter or a steep-cutoff notch filter can only be realized by using digital signal processing techniques, and many adaptive systems are achievable in a practical product only via digital manipulation of signals. In essence, digital representation (0s and 1s) allows voice, audio, image, and video data to be treated the same for error-tolerant digital transmission and storage purposes. As a result, digital processing, and hence digital signal processors (also called DSPs), are expected to play a major role in the next generation of telecommunication infrastructure including 3G (third generation) wireless, cable (cable modems), and telephone lines (digital subscriber line – DSL modems).

The processing of a digital signal can be implemented on various platforms such as a DSP processor, a customized very large scale integrated (VLSI) circuit, or a general-purpose microprocessor. Some of the differences between a DSP and a single function VLSI implementation are as follows:

1. There is a fair amount of application flexibility associated with DSP implementation, since the same DSP hardware can be utilized for different applications. In other words, DSP processors are programmable. This is not the case for a hardwired digital circuit.

2. DSP processors are cost-effective because they are mass-produced and can be used for many applications. A customized VLSI chip normally gets built for a single application and a specific customer.

3. In many situations, new features constitute a software upgrade on a DSP processor not requiring new hardware. In addition, bug fixes are generally easier to make.

4. Often very high sampling rates can be achieved by a customized chip, whereas there are sampling rate limitations associated with DSP chips due to their peripheral constraints and architecture design.

DSP processors share some common characteristics that also separate them from general-purpose microprocessors. Some of these characteristics include the following:

1. They are optimized to cope with repetition or looping of operations common in signal processing algorithms. Relatively speaking, instruction sets of DSPs are smaller and optimized for signal processing operations, such as single-cycle multiplication and accumulation.

2. DSPs allow specialized addressing modes, like indirect and circular addressing. These are efficient addressing mechanisms for implementing many signal processing algorithms.

3. DSPs possess appropriate peripherals that allow efficient input/output (I/O) interfacing to other devices.

4. In DSP processors, it is possible to perform several accesses to memory in a single instruction cycle. In other words, these processors have a relatively high bandwidth between their central processing units (CPUs) and memory.

It should be kept in mind that due to the constant evolving of features being placed on processors, one needs to be cautious of features dividing DSPs and general-purpose microprocessors.

Most of the market share of DSPs belong to real-time, cost-effective, embedded systems, for example, cellular phones, modems, and disk drives. *Real-time* means completing the processing within the allowable or available time between samples. This available time, of course, depends on the application. As illustrated in Figure 1-2, the number of instructions to have an algorithm running in real-time must be less than the number of instructions that can be executed between two consecutive samples. For example, for audio processing operating at 44.1 kHz sampling frequency, or approximately 22.6 μs sampling time interval, the number of instructions must be fewer than nearly 4500, assuming an instruction cycle time of 5 ns. There are two aspects of real-time processing: (a) sampling rate, and (b) system latencies (delays). Typical sampling rates and latencies for several different applications are shown in Table 1-1.

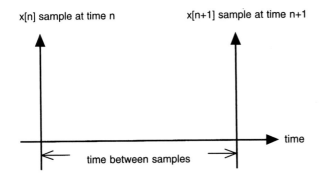

x[n] sample at time n   x[n+1] sample at time n+1

Figure 1-2: Maximum number of instructions to meet real-time = time between samples/instruction cycle time.

time

time between samples

Table 1-1: Typical sampling rates and latencies for select applications.

Application	I/O Sampling Rate	Latency
Instrumentation	1 Hz	*system dependent
Control	> 0.1 kHz	*system dependent
Voice	8 kHz	< 50 ms
Audio	44.1 kHz	*< 50 ms
Video	1–14 MHz	*< 50 ms

*In many cases, one may not need to be concerned with latency, for example, a TV signal is more dependent on synchronization with audio than the latency. In each of these cases, the latency is dependent on the application.

1.1 Examples of DSP Systems

For the reader to appreciate the usefulness of DSPs, several examples of DSP systems currently in use are presented here.

During the past few years, there has been a tremendous growth in the wireless market. Figure 1-3 illustrates a cellular phone wireless communication DSP system. As can be seen from this figure, there are two sets of data converters. On the voice band side, a low sampling rate (for example, 8 kSPS [kilo samples per second]) and a high resolution (for example, 13 bits) converter is used, whereas on the RF modulation side, a relatively high-speed (for example, 20 MSPS) and a low resolution (for example, 8 bits) converter is used. System designers prefer to integrate more func-

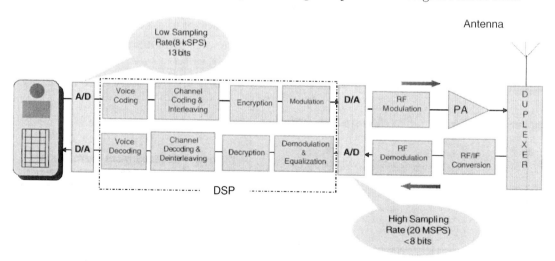

Figure 1-3: Cellular phone wireless communication DSP system.

tionalities in DSP rather than in analog components in order to lower the number of components and hence the overall cost. This strategy of more integration in DSP depends on specifications achievable for low power consumption in portable devices.

In wired communications, various types of modems are used to convert analog/digital signals to digital signals appropriate for error-tolerant transmission over wires or cables. Currently, available modem types include: high-speed voiceband (56 kbps [kilo bits per second]), integrated services digital network (ISDN), DSL, and cable modems. For example, DSL type modems have data rates in the range of 1–52 Mbps. DSL makes use of the existing twisted-pair wires between residential homes and the phone company's central office. For example, the asymmetric version of DSL (ADSL) uses the frequency range 25–138 kHz for upstream and 200 kHz–1.1 MHz for downstream data transmission, without interfering with the existing 0–4 kHz voiceband range. Figure 1-4 shows an ADSL system based on the TI data converters and DSP products. The indicated A/D and D/A converters have a high-speed, high-resolution specification to cope with the multilevel nature of the transmitted signal. The transceiver is a dedicated DSP performing the ADSL modulation/demodulation.

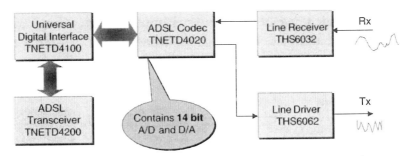

Figure 1-4: TI chipset for ADSL wired communication DSP system.

Considering that communication networks in use today are digital, an analog signal reaching the phone company central office must be conditioned and converted to a digital signal for transmission through the network. Figure 1-5 shows the pulse code modulation (PCM) voiceband codec used in communications networks. As can be seen, a fair amount of the signal processing is done in digital domain by the DSP component.

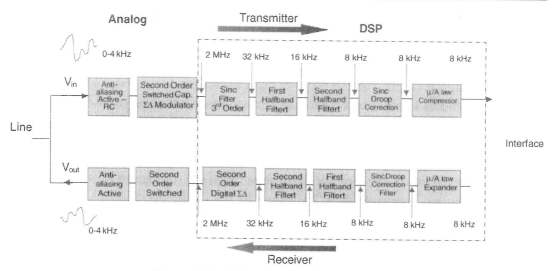

Figure 1-5: PCM voiceband DSP system.

Figure 1-6 shows a gigabit Ethernet DSP system. The analog signal is sent through category-5 twisted-pair wires. Four 8-bit, high-speed A/D converters are used for data conversion. The dynamic range of the converters must be high enough to overcome noise, interference, and attenuation through an Ethernet link. A DSP is then used to do echo cancellation, equalization, and demodulation signal processing.

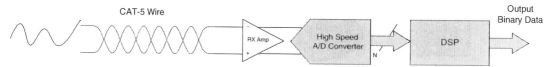

Figure 1-6: Gigabit Ethernet DSP system.

Data stored on a compact disc (CD) or a computer hard drive is in binary format. However, the signal generated by a read head is analog and corrupted by noise and distortion. This demands a fair amount of signal conditioning and filtering after reading data. As shown in Figure 1-7, this is achieved by using a DSP-based hard disk drive system.

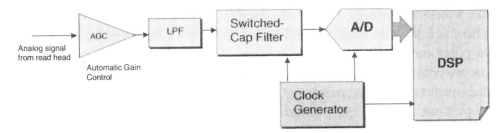

Figure 1-7: Hard disk drive DSP system.

Motor control is another area where DSPs are making an impact. For example, as illustrated in Figure 1-8, DSPs are used to control induction motors via monitoring feedback signals including current, voltage, and position. Such motors are widely used because of their low cost, high reliability, and high efficiency.

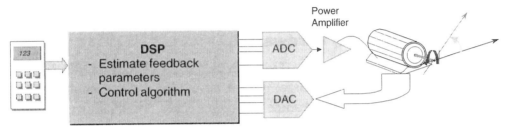

Figure 1-8: Motor control DSP system.

Smart sensors or devices are another example of DSP systems. These sensors are capable of both data acquisition and data processing. An example of such sensors is the airbag activation system in automobiles. Vehicle acceleration is measured by a suspension-mass sensor and converted into a digital signal by an A/D converter. This signal is then processed by a DSP to detect an accident by comparing features of the signal with those of the accident.

1.2 Organization of Chapters

Chapter 2 provides a discussion of the differences and relationships between analog and digital signals. In Chapter 3, an overview of the TMS320C6x architecture is presented. The focus here is placed on the architectural features one needs to be aware of in order to implement algorithms on the C6x processor. In Chapter 4, the C6x software tools are presented, and the steps in taking a source file to an executable file are discussed. Lab 1 in Chapter 4 provides a hands-on approach for becoming familiar with the Code Composer Studio™ integrated development environment.

Chapter 5 presents the concept of interrupt data processing. Lab 2 in Chapter 5 shows how to sample an analog signal in real-time on a C6x target board. In Chapter 6, fixed-point and floating-point number representations are discussed and their differences are pointed out. Lab 3 in Chapter 6 gives suggestions on how one may cope with the overflow or scaling problem. Code efficiency issues appear in Chapter 7, in which optimization techniques, as well as linear assembly and hand-coded pipelined assembly, are discussed. Lab 4 in Chapter 7 covers Finite Impulse Response (FIR) filtering while deploying various optimization techniques. Chapter 8 covers circular buffering. Lab 5 in Chapter 8 shows how circular buffering is used to perform adaptive filtering. Frame processing is covered in Chapter 9. Lab 6 in Chapter 9 provides an example of frame processing involving fast Fourier transform (FFT) implementation and the use of direct memory access (DMA). Chapter 10 and Labs 7 and 8 address the DSP/BIOS real-time analysis and scheduling features of Code Composer Studio. Finally, four project examples are presented in Chapter 11.

1.3 Required Software/Hardware

The software tool needed to generate TMS320C6x executable files is called Code Composer Studio (CCS). CCS incorporates the assembler, linker, compiler, simulator, and debugger utilities. In the absence of a target board, which allows one to run an executable file on an actual C6x processor, the simulator can be used to verify code functionality by using data already stored in a datafile. However, when using the simulator, an Interrupt Service Routine (ISR) cannot be used to read in signal samples from a signal source. To be able to process signals in real-time on an actual C6x processor, a DSP Starter Kit (DSK) or an EValuation Module (EVM) board is needed for code development. The recommended testing equipment are a function generator, oscilloscope, microphone, boom box, and cables with audio jacks.

A DSK board can easily be connected to a PC host through its parallel or USB port. The signal interfacing with the DSK board is done through its two standard audio jacks. An EVM board needs to be installed in a full-length PCI slot inside a PC host. Refer to the *TI TMS320C6x Evaluation Module Reference Guide* [1] for the installation details. The signal interfacing with the EVM board is done through its three standard audio jacks.

For performing the labs, familiarity with C is assumed. The accompanying CD-ROM includes the lab codes for the following C6x DSP target boards: C6711 DSK, C6416/C6713 DSK, and C6701/C6201 EVM.

Bibliography

[1] Texas Instruments, *TMS320C6201/6701 Evaluation Module Reference Guide*, Literature ID# SPRU 269F, 2002.

CHAPTER 2

Analog-to-Digital Signal Conversion

The process of analog-to-digital signal conversion consists of converting a continuous time and amplitude signal into discrete time and amplitude values. Sampling and quantization constitute the steps needed to achieve analog-to-digital signal conversion. To minimize any loss of information that may occur as a result of this conversion, it is important to understand the underlying principles behind sampling and quantization.

2.1 Sampling

Sampling is the process of generating discrete time samples from an analog signal. First, it is helpful to see the relationship between analog and digital frequencies. Let us consider an analog sinusoidal signal $x(t) = A\cos(\omega t + \phi)$. Sampling this signal at $t = nT_s$, with the sampling time interval of T_s, generates the discrete time signal

$$x[n] = A\cos(\omega n T_s + \phi) = A\cos(\theta n + \phi), \qquad n = 0,1,2,\ldots, \qquad (2.1)$$

where $\theta = \omega T_s = \dfrac{2\pi f}{f_s}$ denotes digital frequency with units radians (as compared to analog frequency ω with units radians/sec).

The difference between analog and digital frequencies is more evident by observing that the same discrete time signal is obtained for different continuous time signals if the product ωT_s remains the same. (An example is shown in Figure 2-1.) Likewise, different discrete time signals are obtained for the same analog or continuous time signal when the sampling frequency is changed. (An example is shown in Figure 2-2.) In other words, both the frequency of an analog signal and the sampling frequency define the frequency of the corresponding digital signal.

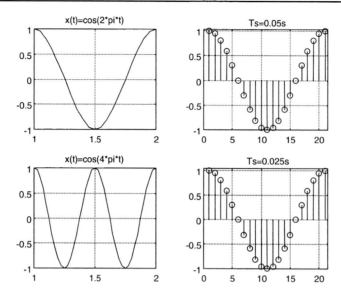

Figure 2-1: Different sampling of two different analog signals leading to the same digital signal.

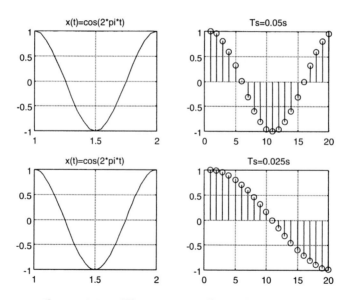

Figure 2-2: Different sampling of the same analog signal leading to two different digital signals.

It helps to understand the constraints associated with the above sampling process by examining signals in frequency domain. The Fourier transform pairs in analog and digital domains are given by

Fourier transform pair for analog signals

$$\begin{cases} X(j\omega) = \int_{-\infty}^{\infty} x(t)e^{-j\omega t} dt \\ x(t) = \dfrac{1}{2\pi} \int_{-\infty}^{\infty} X(j\omega)e^{j\omega t} d\omega \end{cases}$$

$$(2.2)$$

Fourier transform pair for discrete signals

$$\begin{cases} X(e^{j\theta}) = \sum_{n=-\infty}^{\infty} x[n]e^{-jn\theta}, \ \theta = \omega T_s \\ x[n] = \dfrac{1}{2\pi} \int_{-\pi}^{\pi} X(e^{j\theta})e^{jn\theta} d\theta \end{cases}$$

$$(2.3)$$

Analog Waveform

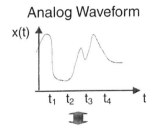

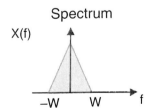

Discrete Signal

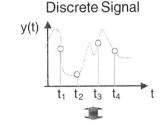

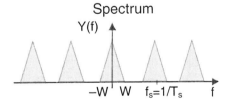

Figure 2-3: (a) Fourier transform of a continuous-time signal, and (b) its discrete time version.

As illustrated in Figure 2-3, when an analog signal with a maximum frequency of f_{max} (or bandwidth of W) is sampled at a rate of $T_s = \dfrac{1}{f_s}$, its corresponding frequency response is repeated every 2π radians, or f_s. In other words, Fourier transform in digital domain becomes a periodic version of Fourier transform in analog domain. That is why, for discrete signals, we are only interested in the frequency range $0 - f_s/2$.

Therefore, in order to avoid any aliasing or distortion of the frequency content of the discrete signal, and hence to be able to recover or reconstruct the frequency content of the original analog signal, we must have $f_s \geq 2f_{max}$. This is known as the *Nyquist rate*; that is, the sampling frequency should be at least twice the highest frequency in the signal. Normally, before any digital manipulation, a frontend antialiasing analog lowpass filter is used to limit the highest frequency of the analog signal.

Figure 2-4 shows the Fourier transform of a sampled sinusoid with a frequency of f_o. As can be seen, there is only one frequency component at f_o. The aliasing problem can be further illustrated by considering an undersampled sinusoid as depicted in Figure 2-5. In this figure, a 1 kHz sinusoid is sampled at $f_s = 0.8$ kHz, which is less than the Nyquist rate. The dashed-line signal is a 200 Hz sinusoid passing through the same sample points. Thus, at this sampling frequency, the output of an A/D converter would be the same if either of the sinusoids were the input signal. On the other hand, oversampling a signal provides a richer description than that of the same signal sampled at the Nyquist rate.

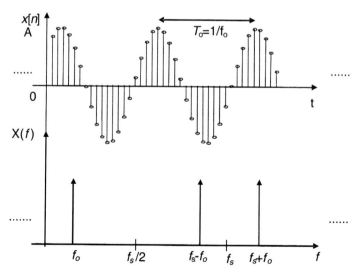

Figure 2-4: Fourier transform of a sampled sinusoidal signal.

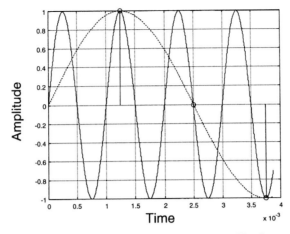

Figure 2-5: Ambiguity caused by aliasing.

2.1.1 Fast Fourier Transform

Fourier transform of discrete signals is continuous over the frequency range $0 - f_s/2$. Thus, from a computational standpoint, this transform is not suitable to use. In practice, discrete Fourier transform (DFT) is used in place of Fourier transform. DFT is the equivalent of Fourier series in analog domain. However, it should be remembered that DFT and Fourier series pairs are defined for periodic signals. These transform pairs are expressed as

Fourier series for periodic analog signals
$$\begin{cases} X_k = \dfrac{1}{T}\displaystyle\int_{-T/2}^{T/2} x(t)e^{-j\omega_0 kt}dt \\[2ex] x(t) = \displaystyle\sum_{k=-\infty}^{\infty} X_k e^{j\omega_0 kt} \\[2ex] \text{where } T \text{ denotes period and} \\ \omega_0 \text{ fundamental frequency.} \end{cases}$$

$$(2.4)$$

Discrete Fourier Transform (DFT) for periodic discrete signals
$$\begin{cases} X[k] = \displaystyle\sum_{n=0}^{N_S-1} x[n]e^{-j\frac{2\pi}{N_S}nk} , k = 0,1,...,N_s -1 \\[2ex] x[n] = \dfrac{1}{N_S}\displaystyle\sum_{k=0}^{N_S-1} X[k]e^{j\frac{2\pi}{N_S}nk} , n = 0,1,...,N_s -1 \end{cases}$$

$$(2.5)$$

15

Hence, when computing DFT, it is required to assume periodicity with a period of N_s samples. Figure 2-6 illustrates a sampled sinusoid which is no longer periodic. In order to make sure that the sampled version remains periodic, the analog frequency should satisfy this condition [1]

$$f_o = \frac{m}{N_s} f_s,$$ (2.6)

where m denotes number of cycles over which DFT is computed.

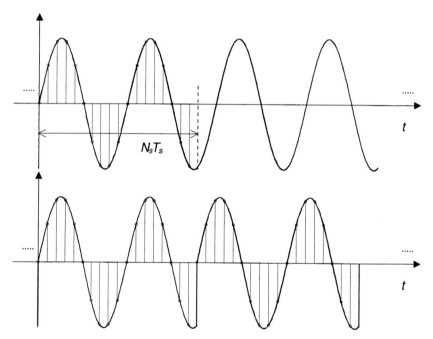

Figure 2-6: Periodicity condition of sampling.

The computational complexity (number of additions and multiplications) of DFT is reduced from N_s^2 to $N_s log N_s$ by using fast Fourier transform (FFT) algorithms. In these algorithms, N_s is considered to be a power of two. Figure 2-7 shows the effect of the periodicity constraint on the FFT computation. In this figure, the FFTs of two sinusoids with frequencies of 250 Hz and 251 Hz are shown. The amplitudes of the sinusoids are unity. Although there is only a 1 Hz difference between the sinusoids, the FFT outcomes are significantly different due to improper sampling.

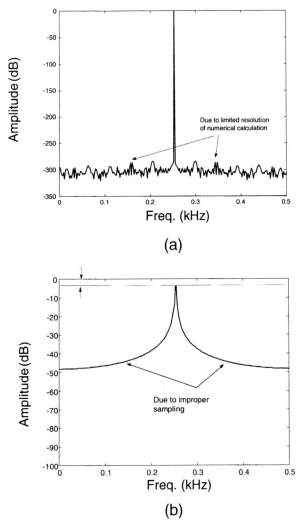

(a)

(b)

Figure 2-7: FFTs of a 250 and a 251 Hz sinusoids.

2.1.2 Amplitude Statistics

An important property used in signal analysis is amplitude statistics. This statistics reflects the probability density function (pdf) associated with amplitudes of a randomly sampled signal. In other words, this pdf shows the histogram of sample points if the signal is sampled with infinitesimal sampling period. For example, for a sinewave $x(t) = a\sin(2\pi\ f_o t)$, its amplitude pdf is given by

$$f(x) = \frac{1}{\pi\sqrt{a^2 - x^2}}, \qquad |x| < a$$

(2.7)

This PDF is illustrated in Figure 2-8.

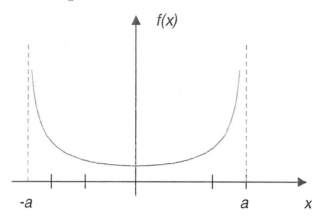

Figure 2-8: Amplitude PDF of sinewave.

Consider a sinewave with f_o = 1 kHz, N_s = 80, and m = 10. As shown in Figure 2-9, the amplitude histogram is not correct due to improper sampling or by repeatedly sampling the same level. In order to avoid this sampling outcome and obtain a proper amplitude statistics, the number of cycles m and the number of samples N_s must be mutually prime. Figure 2-10 shows the amplitude histogram for m = 13.

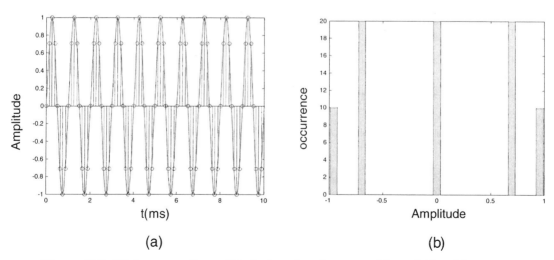

(a)   (b)

Figure 2-9: Histogram of amplitude levels when m = 10 and N_s = 80 are not mutually prime: (a) sampled signal, and (b) histogram of sample points.

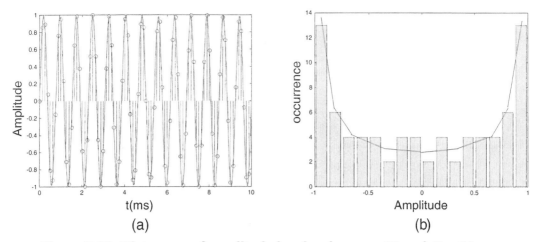

Figure 2-10: Histogram of amplitude levels when *m* = 13 and *N_s* = 80 are mutually prime: (a) sampled signal, and (b) histogram of sample points.

2.1.3 Harmonics of Distorted Sinewaves

Linear circuits, in general, are not perfectly linear, leading to distortion of output signals. Circuit designers measure this distortion by using a sinewave input. A distorted sinewave, as indicated by Fourier series, contains harmonics. In analog domain, the locations of harmonics on the frequency axis are easy to predict. These locations are at kf_o where k denotes harmonic index. However, as a result of sampling, the locations of harmonics are not so easy to predict because of aliasing. The Nyquist rate condition is usually held for the fundamental frequency. Consequently, the sampling frequency may not be sufficient for higher harmonics. Knowledge of the locations of harmonics is of great importance in the interpretation of FFT results, especially for diagnostic purposes. Fig. 2-11 shows the effect of sampling on harmonics index. It is seen that the sampling of a distorted sinewave results in consecutive folding of harmonics between 0 and $f_s/2$. Figure 2-12 shows the FFT result of a distorted sinewave with f_o = 1.5 kHz and f_s = 10 kHz.

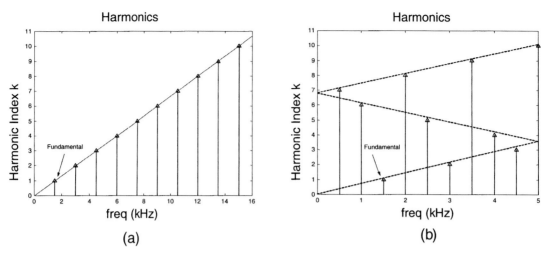

Figure 2-11: Effect of sampling on harmonic index; $f_o = 1.5$ **kHz with 10 harmonics and** $f_s = 10$ **kHz: (a) before sampling, and (b) after sampling.**

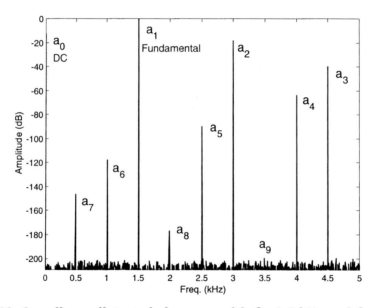

Figure 2-12: Sampling a distorted sinewave with $f_o = 1.5$ **kHz and** $f_s = 10$ **kHz.**

2.2 Quantization

An A/D converter has a finite number of bits (or resolution). As a result, continuous amplitude values get represented or approximated by discrete amplitude levels. The process of converting continuous into discrete amplitude levels is called quantization. This approximation leads to an error called quantization noise. The input/output characteristic of a 3-bit A/D converter is shown in Figure 2-13 to see how analog voltage values get approximated by discrete voltage levels.

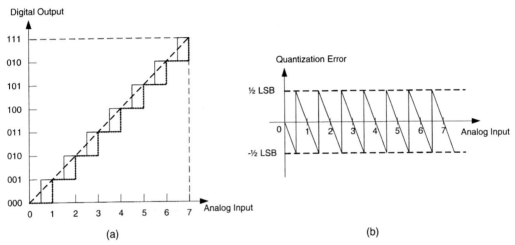

(a)   (b)

**Figure 2-13: Characteristic of a 3-bit A/D converter:
(a) input/output static transfer function, and (b) additive quantization noise.**

The quantization interval depends on the number of quantization or resolution level, as illustrated in Figure 2-14. Clearly the amount of quantization noise generated by an A/D converter depends on the size of quantization interval. More quantization bits translate into a narrower quantization interval and hence into a lower amount of quantization noise.

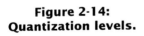
**Figure 2-14:
Quantization levels.**

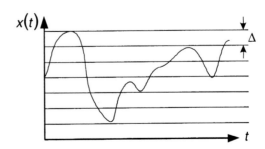

To avoid saturation or out-of-range distortion, the input voltage must be between V_{ref-} and V_{ref+}. The full-scale (FS) voltage or V_{ref} is defined as

$$V_{FS} = V_{ref} = V_{ref+} - V_{ref-} \qquad (2.8)$$

and 1 least significant bit (LSB) is given by

$$1\ \text{LSB} = \Delta = \frac{V_{ref}}{2^N}, \qquad (2.9)$$

where N is the number of bits of the A/D converter. Table 2-1 lists 1 LSB in volts for different numbers of bits and reference voltages. It is interesting to note that a couple of microvolts, which is 1 LSB in a high-resolution A/D converter, can be generated by a dozen electrons in a 1 pF capacitor!

Table 2-1: LSB of A/D converter.

N	8	10	12	14	16	20
Vref = 5 V	19.5 mV	4.9 mV	1.2 mV	305 µV	76 µV	4.8 µV
Vref = 3 V	11.7 mV	2.9 mV	732 µV	183 µV	45.8 µV	2.8 µV
Vref = 1.8 V	7.0 mV	1.7 mV	439 µV	110 µV	27.5 µV	1.7 µV

Usually, it is assumed that quantization noise is signal independent and is uniformly distributed over −0.5 LSB and 0.5 LSB. Figure 2-15 shows the quantization noise of an analog signal quantized by a 3-bit A/D converter. It is seen that, although the histogram of the quantization noise is not exactly uniform, it is reasonable to accept the uniformity assumption.

Figure 2.16 shows the FFT of a sinewave before and after the digitization process. The input sinewave is at 250 Hz, with unity amplitude, f_s = 1 kHz, and N_s = 512. It is seen that the quantization error raises the noise level.

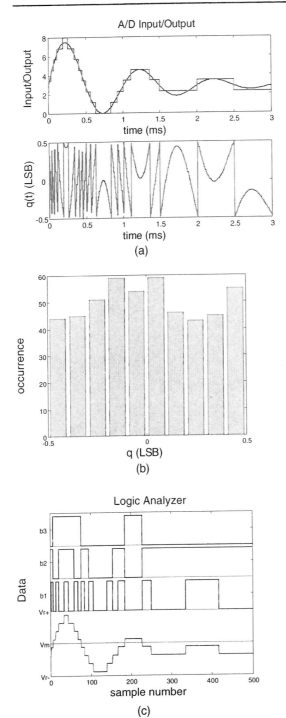

Figure 2-15: Quantization of an analog signal by a 3-bit A/D converter:

(a) output signal and quantization error, (b) histogram of quantization error, and (c) bit stream.

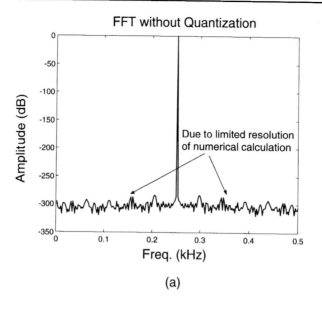

(a)

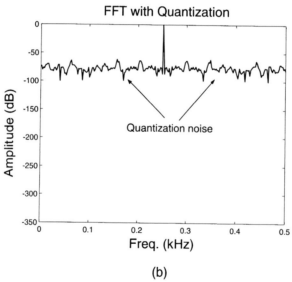

(b)

Figure 2-16: Sinewave before and after digitization, $f_o = 250$ Hz, $f_s = 1$ kHz, $N_s = 512$, $N = 8$-bit: (a) FFT before digitization, and (b) FFT after digitization.

2.2.1 Signal-to-Noise Ratio

Resolution is the term used to describe the minimum resolvable signal level by an A/D converter. The fundamental limit of an A/D converter is governed by quantization noise, which is caused by the A/D converter's finite resolution. If the output digital word consists of N bits, the minimum step that the converter can resolve is 1 LSB. If we assume quantization error, n_q, is a random variable uniformly distributed and independent of the input signal, then we have

$$\delta_q^2 = E\left[n_q^2\right] = \frac{1}{\Delta}\int_{-\frac{\Delta}{2}}^{\frac{\Delta}{2}} n_q^2 dn_q = \frac{\Delta^2}{12}, \tag{2.10}$$

where δ_q^2 indicates quantization noise variance. For a sinusoidal input signal having an amplitude of A_m, an ideal A/D converter has a Signal-to-Noise Ratio (SNR) of

$$10\log\frac{P_S}{P_n} = 10\log\frac{\left(A_m\right)^2/2}{\frac{1}{12}\left(V_{ref}/2^N\right)^2}, \tag{2.11}$$

where P_S and P_n denote signal and noise power, respectively. It is observed that the quantization SNR is a function of amplitude. The maximum SNR can thus be written as

$$\text{SNR}_{max} = 10\log\frac{\left(V_{ref}/2\right)^2/2}{\frac{1}{12}\left(V_{ref}/2^N\right)^2} = 10\log\frac{3}{2}2^{2N} = 6.02N + 1.76 \;\; (\text{dB}) \tag{2.12}$$

For instance, an ideal 16-bit A/D converter has a maximum SNR of about 97.8 dB. Quantization noise decreases by 6 dB for each additional bit.

Figure 2-17 shows the SNR of an 8-bit A/D converter as a function of the input amplitude. The maximum occurs when the input sinewave amplitude is equal to one half of the full-scale voltage (scaled to 0 dB).

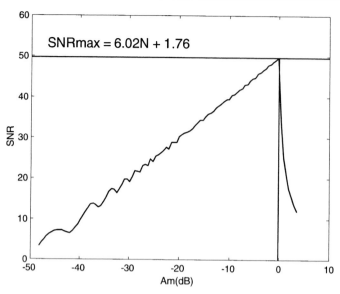

Figure 2-17: Signal-to-noise ratio of an ideal 8-bit A/D converter.

To better understand the quantization effect, let us assume that the signal is zero mean Gaussian with $V_{ref} = K\sigma_x$, where σ_x denotes standard deviation of the signal. By substituting for V_{ref} and σ_x in (2.11), we obtain

$$\text{SNR} = 10\log\frac{\sigma_x^2}{\sigma_q^2} \cong 6N + 10.8 - 20\log_{10} K \tag{2.13}$$

As an example, consider $K = 4$. The probability of signal samples falling in the $4\sigma_x$ range is 0.954. This means that out of 1,000 samples, 954 samples fall in this range on average. In other words, 46 out of 1,000 samples fall outside the indicated range and hence get represented by the maximum or minimum allowable value.

If the signal is scaled by α, the corresponding signal variance changes to $\alpha^2\sigma_x^2$. Hence, the SNR changes to

$$\text{SNR} \cong 6N + 10.8 - 20\log_{10} K + 20\log_{10} \alpha \tag{2.14}$$

It is important to note that when we perform fractional arithmetic (discussed later in Chapter 6), α is scaled to be less than 1, leading to a lower signal-to-noise ratio. This indicates that quantization noise should be kept in mind when scaling down the input signal. In other words, scaling down to achieve fractional representation cannot be done indefinitely, since, as a result, the signal would get buried in quantization noise.

The interested reader is referred to [2] for more elaborate analysis of quantization noise. For example, for a linear time-invariant system such as a FIR or an IIR filter, it can be shown that the noise variance σ_o^2 at the output of the system, caused by the input quantization noise, is given by

$$\sigma_o^2 = \sigma_q^2 \sum_n h^2[n] \tag{2.15}$$

where h denotes the unit sample response.

2.3 Signal Reconstruction

So far, we have examined the forward process of sampling. It is also important to understand the inverse process of signal reconstruction from samples. According to the Nyquist theorem, an analog signal v_a can be reconstructed from its samples by using the following formula:

$$v_a(t) = \sum_{k=-\infty}^{\infty} v_a(kT_s) \left[\operatorname{sinc}\left(\frac{t - kT_s}{T_s} \right) \right] \tag{2.16}$$

One can see that the reconstruction is based on the interpolation of shifted sinc functions. Figure 2-18 illustrates the reconstruction of a sinewave from its samples.

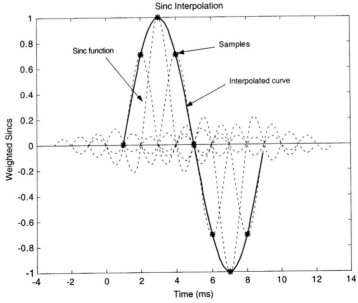

Figure 2-18: Reconstruction of an analog sinewave based on its samples
$A_m = 1$, $f_s = 2$ Hz, and $f_s = 10$ kHz.

It is very difficult to generate sinc functions by electronic circuitry. That is why, in practice, an approximation of sinc function is used. Figure 2-19 shows an approximation of a sinc function by a pulse, which is easy to realize in electronic circuitry. In fact, the well-known sample and hold circuit performs this approximation. The final stage of a D/A converter is the sample and hold circuit. The transfer function of a D/A converter is

$$H(j\omega) = \frac{1}{j\omega} - \frac{1}{j\omega}e^{-j\omega T_s} = \frac{\sin(\omega T_s/2)}{\omega T_s/2}e^{-j\omega T_s/2} = \text{sinc}(f/f_s)e^{-j\pi \, f/f_s}$$

(2.17)

Both the time and frequency domain response of a D/A converter are shown in Figure 2-20.

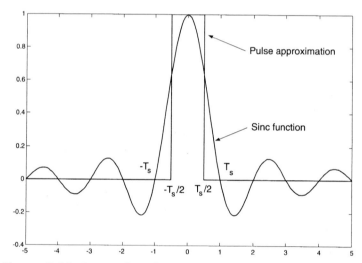

Figure 2-19: Approximation of a sinc function by a pulse.

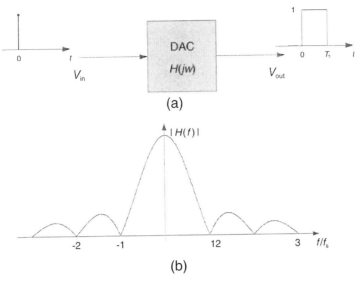

(a)

(b)

Figure 2-20: D/A converter: (a) time-domain, and (b) frequency response.

Sample and hold circuits in D/A converters have two inherent non-idealities. First, as illustrated in Figure 2-21, the gain in the desired central band is not constant. It is possible to compensate for this non-ideality by using an inverse filter as part of the DSP component. Another solution is to increase sampling frequency, which results in a narrower relative signal bandwidth. The second non-ideality is caused by the presence of high-frequency replica of the signal spectrum, which can be removed by using a lowpass filter. These solutions are illustrated in Figure 2-22.

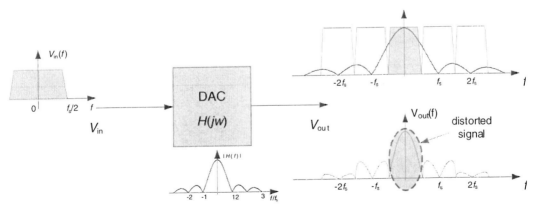

Figure 2-21: Non-idealities of a D/A converter.

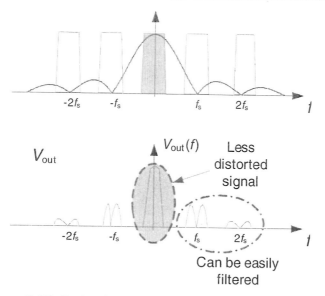

Figure 2-22: Reduction of frequency distortion of a D/A converter by increasing sampling frequency.

Bibliography

[1] J. Proakis and D. Manolakis, *Digital Signal Processing:Principles, Algorithms, and Applications*, Prentice-Hall, 1996.

[2] S. Mitra, *Digital Signal Processing: A Computer-Based Approach*, McGraw Hill, 1998.

[3] B. Razavi, *Principles of Data Conversion System Design*, IEEE Press, 1995.

CHAPTER 3

TMS320C6x Architecture

The choice of a DSP processor to implement an algorithm in real-time is application dependent. There are many factors that influence this choice. These factors include cost, performance, power consumption, ease-of-use, time-to-market, and integration/ interfacing capabilities.

The family of TMS320C6x processors, manufactured by Texas Instruments™, are built to deliver speed. They are designed for million instructions per second (MIPS) intensive applications such as 3G wireless, DSL/cable modems, and digital imaging. Table 3-1 provides a list of currently available fixed-point and floating-point C6x processors at the time of this writing. As can be seen from this table, instruction cycle time, speed, power consumption, memory, peripherals, packaging, and cost specifications vary for different products in this family. For example, the fixed-point C6416-600 version operates at 600 MHz (1.67 ns cycle time), delivering a peak performance of 4800 MIPS. The floating-point C6713-225 version operates at 225 MHz (10 ns cycle time), delivering a peak performance of 1350 MIPS. Figure 3-1 illustrates the processing power of C6x by showing a speed benchmarking comparison with some other common DSP processors.

Table 3-1: Sample C6x DSP product specifications (year 2003).[†]

Device	RAM(Bytes) Data / Prog	McBSP	(E)DMA	COM	Timers	MHz	Cycles (ns)	MIPS	Typical Activity- Total Internal Power (W) (Full Device Speed)	Voltage (V) Core, I/O	Packaging
TMS320C6711-100	4K/4K/64K	2	16	HPI/16	2	100	10	600	1.1	1.8, 3.3	256 BGA, 27 mm
TMS320C6711-150	4K/4K/64K	2	16	HPI/16	2	150	6.7	900	1.1	1.8, 3.3	256 BGA, 27 mm
TMS320C6713-200	4K/4K/256K	2	16	HPI/16	2	200	5	1200	1.0	1.2, 3.3	208 TQFP, 28 mm
TMS320C6713-225	4K/4K/256K	2	16	HPI/16	2	225	4.4	1350	1.2	1.26, 3.3	272 BGA, 27 mm
TMS320C6701-150	64K/64K	2	4	HPI/16	2	150	6.7	900	1.3	1.8, 3.3	352 BGA, 35 mm
TMS320C6701-167	64K/64K	2	4	HPI/16	2	167	6	1000	1.4	1.9, 3.3	352 BGA, 35 mm
TMS320C6416-500	16K/16K/1M	2+UTOPIA*	64	PCI/HPI 32/16	3	500	2	4000	0.64	1.2, 3.3	532 BGA, 23 mm
TMS320C6416-600	16K/16K/1M	2+UTOPIA*	64	PCI/HPI 32/16	3	600	1.67	4800	1.06	1.4, 3.3	532 BGA, 23 mm

* UTOPIA pins muxed with a third McBSP.

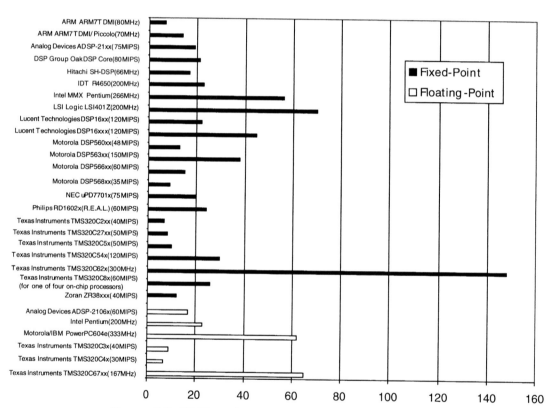

Figure 3-1: BDTImark™ DSP Speed Metric benchmark by Berkeley Design Technology, Inc.[1]

---

[1] The BDTImark is a summary measure of DSP speed, distilled from a suite of DSP benchmarks developed and independently verified by Berkeley Design Technology, Inc. A higher BDTImark score indicates a faster processor. For a complete description of the BDTImark and underlying benchmarking methodology, as well as additional BDTImark scores, refer to http://www.bdti.com. © 2000 Berkeley Design Technology, Inc.

Figure 3-2 shows the block diagrams of the generic C6x, C64x, and C6211/C6711 architectures. The C6x CPU consists of eight functional units divided into two sides: (A) and (B). Each side has a so-called .M unit (used for multiplication operation), a .L unit (used for logical and arithmetic operations), a .S unit (used for branch, bit manipulation and arithmetic operations), and a .D unit (used for loading, storing and arithmetic operations). Some instructions such as ADD can be done by more than one unit. There are sixteen 32-bit registers associated with each side. Interaction with the CPU must be done through these registers. A listing of the C6x instructions, as divided by the four functional units, appears in Appendix A (Quick Reference Guide). These instructions are fully discussed in the *TI TMS320C6000 CPU and Instruction Set Reference Guide* [1].

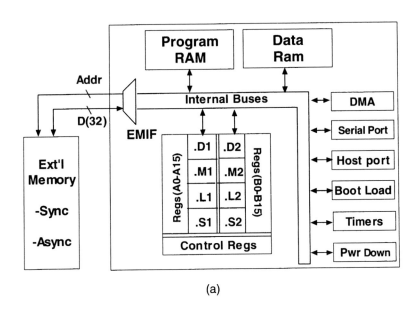

(a)

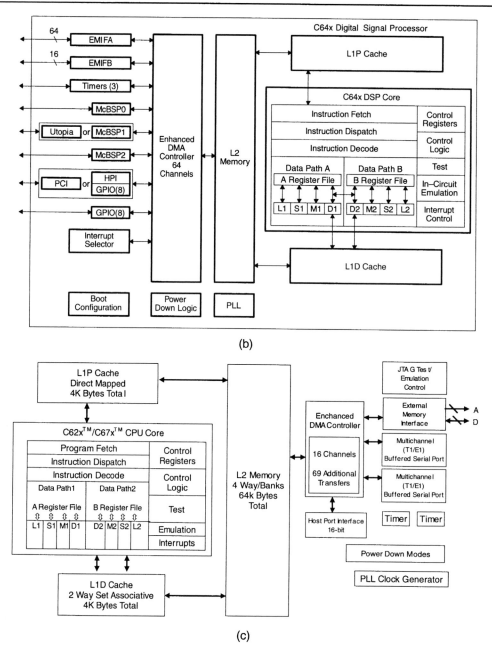

Figure 3-2:
(a) Generic C6x architecture,
(b) C64x architecture, and
(c) C6211/C6711 architecture.[†]

As shown in Figure 3-3, the internal buses consist of a 32-bit program address bus, a 256-bit program data bus accommodating eight 32-bit instructions, two 32-bit data address buses (DA1 and DA2), two 32-bit (64-bit for C64 version) load data buses (LD1 and LD2), and two 32-bit (64-bit for the floating-point version) store data buses (ST1 and ST2). In addition, there are a 32-bit DMA data and a 32-bit DMA address bus. The off-chip, or external, memory is accessed through a 20-bit address bus and a 32-bit data bus.

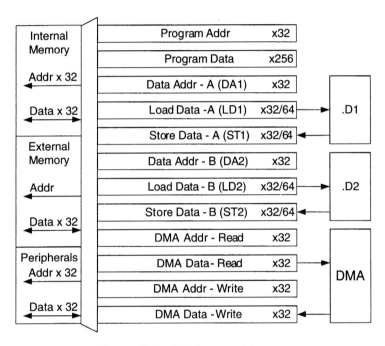

Figure 3-3: C6x internal buses.

The peripherals on a typical C6x processor include External Memory Interface (EMIF), DMA, Boot Loader, Multichannel Buffered Serial Port (McBSP), Host Port Interface (HPI), Timer, and Power Down unit. EMIF provides the necessary timing for accessing external memory. DMA allows the movement of data from one place in memory to another place without interfering with the CPU operation. Boot Loader boots the loading of code from off-chip memory or HPI to internal memory. McBSP provides a high-speed multichannel serial communication link. HPI allows a host to access internal memory. Timer provides two 32-bit counters. Power Down unit is used to save power for durations when the CPU is inactive.

3.1 CPU Operation (Dot Product Example)

As shown in Figure 3-2, the C6x CPU is divided into two data paths, data path A (or 1), and data path B (or 2). An effective way to understand the CPU operation is by going through an example. Figure 3-4 shows the assembly code for a 40-point dot product y between two vectors **a** and **x**, $y = \sum_{n=1}^{40} a_n * x_n$. This code appears in the

TI Technical Training Notes on TMS320C6x DSP [2]. At this point, it is worth mentioning that the assembler is not case sensitive (i.e., instructions and registers can be written in lower or uppercase).

Label	Instruction		Operands	Comment
☐	MVK	.S1	a,A5	;move address of a
☐	MVKH	.S1	a,A5	;into register A5
☐	MVK	.S1	x,A6	;move address of x
☐	MVKH	.S1	x,A6	;into register A6
☐	MVK	.S1	y,A7	;move address of y
☐	MVKH	.S1	y,A7	;into register A7
●	MVK	.S1	40,A2	;A2=40, loop counter
△ loop:	LDH	.D1	*A5++,A0	;A0=a_n
△	LDH	.D1	*A6++,A1	;A1=x_n
	MPY	.M1	A0,A1,A3	;A3=a_n*x_n, product
	ADD	.L1	A3,A4,A4	;y=y+A3
●	SUB	.L1	A2,1,A2	;decrement loop counter
● [A2]	B	.S1	loop	;if A2≠0, branch to loop
△	STH	.D1	A4,*A7	;*A7=y

functional unit ⟋   ⟍ data path: 1 indicates A side and 2, B side

Figure 3-4: Dot product assembly code.

The registers assigned to a_n, x_n, loop counter, product, y, &a[n] (address of a_n), &x[n] (address of x_n), and &y[n] (address of y_n) are shown in Figure 3-5. In this example, only the A side functional units and registers are used.

A loop is created by the instructions indicated by ●'s. First, a loop counter is set up by using the move constant instruction MVK. This instruction uses the .S1 unit to place the constant 40 in register A2. The beginning of the loop is indicated by the label loop and the end by a subtract instruction SUB to decrement the loop counter followed by a branch instruction B to return to loop.

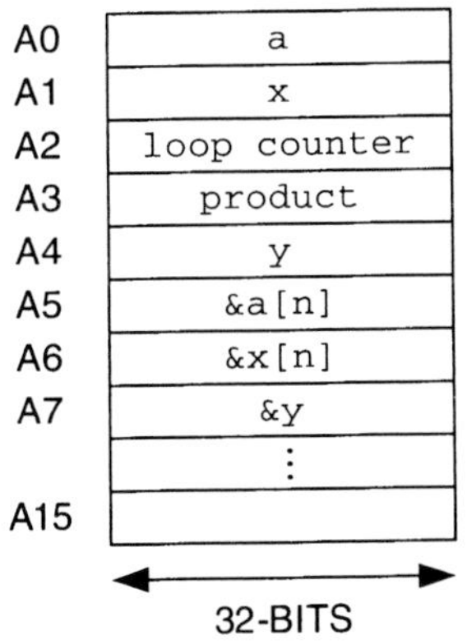

A0	a
A1	x
A2	loop counter
A3	product
A4	y
A5	&a[n]
A6	&x[n]
A7	&y
	⋮
A15	

←——— 32-BITS ———→

Figure 3-5: A-side registers.

The subtraction is performed by the .L1 unit and branching by the .S1 unit. The brackets as part of the branch instruction indicate that this is a conditional instruction. All the C6x's instructions can be made conditional based on a zero or nonzero value in one of the registers: A1, A2, B0, B1, and B2. The syntax [A2] means "execute the instruction if A2 ≠ 0", and [!A2] means "execute the instruction if A2 = 0". As a result of these instructions, the loop is repeated 40 times.

Considering that the interaction with the functional units is done through the A-side registers, these registers must be set up in order to start the loop. The instructions labeled by □'s indicate the necessary instructions for doing so. MVK and MVKH are used to load the address of a_n, x_n, and y into the registers A5, A6, and A7, respectively. These instructions must be done in the order indicated to load the lower 16 bits of the full 32-bit address first, followed by the upper 16 bits. These registers are used as pointers to load a_n, x_n into the A0, A1 registers and store y from the A4 register (instructions labeled by Δ). The C programming language notation * is used to indicate a register is being used as a pointer. Depending on the datatype, any of the following loading instructions can be used: bytes (8-bit) LDB, halfwords (16-bit) LDH, or words (32-bit) LDW. Here, the data is assumed to be halfwords. The loading/ storing is done by the .D1 unit, since .D units are the only units capable of interacting with data memory.

Note that the pointers A5 and A6 need to be post-incremented (C notation), so that they point to the next values for the next iteration of the loop. When registers are used as pointers, there are several ways to perform pointer arithmetic. These include pre- and post-increment/decrement options by some displacement amount, where the pointer is modified before or after it is used (for example, *++A1[disp] and

*A1++[disp]). In addition, a pre-offset option can be performed with no modification of the pointer (for example, *+A1[disp]). Displacement within brackets specifies the number of data elements (depending on the datatype), whereas displacement in parentheses specifies the number of bytes. These pointer offset options are listed in Figure 3-6 together with some examples.

Syntax	Description	Pointer Modified
*R	Pointer	No
*+R[disp]	+Pre-offset	No
*-R[disp]	−Pre-offset	No
*++R[disp]	Pre-increment	Yes
*--R[disp]	Pre-decrement	Yes
*R++[disp]	Post-increment	Yes
*R--[disp]	Post-decrement	Yes

[disp] specifies # elements - size in W, H, or B
(disp) specifies # bytes

(a)

```
A0 [  8  ]                Examples                        Results
A3 [  4  ]

 0  [ FEED ]   1. LDH   *A0--[A3],A5      ;A0=0    A5=0004
 2  [ 00B1 ]
 4  [ 002E ]   2. LDH   *++A3(3),A5       ;A3=6    A5=0033
 6  [ 0033 ]
 8  [ 0004 ]   3. LDB   *+A0[A0],A5       ;A0=8    A5=7A
 A  [ 0095 ]
 C  [ 006C ]   4. LDH   *--A3[0],A5       ;A3=4    A5=002E
 E  [ 0070 ]
10  [ FF7A ]   5. LDB   *-A0[3],A5        ;A0=8    A5=00
```

(b)

**Figure 3-6: (a) Pointer offsets, (b) pointer examples
(note: instructions are independent, not sequential).**[†]

Finally, the instructions MPY and ADD within the loop perform the dot product operation. The instruction MPY is done by the .M1 unit and ADD by the .L1 unit. It should be mentioned that the above code as is will not run properly on the C6x because of its pipelined CPU, which is discussed next.

3.2 Pipelined CPU

In general, it takes several steps to perform an instruction. Basically, these steps are fetching, decoding, and execution. If these steps are done serially, not all of the resources on the processor, such as multiple buses or functional units, are fully utilized. In order to increase throughput, DSP CPUs are designed to be *pipelined*. This means that the foregoing steps are carried out simultaneously. Figure 3-7 illustrates the difference in processing time for three instructions executed on a serial or non-pipelined and a pipelined CPU. As can be seen, a pipelined CPU requires fewer clock cycles to complete the same number of instructions.

	Clock Cycles								
CPU Type	1	2	3	4	5	6	7	8	9
Non-Pipelined	F_1	D_1	E_1	F_2	D_2	E_2	F_3	D_3	E_3

Pipelined

F_1	D_1	E_1		
	F_2	D_2	E_2	
		F_3	D_3	E_3

F_x = fetching of instruction x
D_x = decoding of instruction x
E_x = execution of instruction x

Figure 3-7: Pipelined vs. non-pipelined CPU.[†]

On the C6x processor, fetching consists of four phases, each requiring a clock cycle. These include generate fetch address (denoted by F1), send address to memory (F2), wait for data (F3), and read opcode from memory (F4). Decoding consists of two phases, each requiring a clock cycle. These are dispatching to appropriate functional units (denoted by D1), and decoding (D2). Due to the delays associated with the instructions multiply (MPY – 1 delay), load (LDx – 4 delays), and branch (B – 5 delays), the execution step may consist of up to six phases (denoted by E1 through E6), accommodating a maximum of 5 delays. Hence, as shown in Figure 3-8, the F step consists of four, the D step of two, and the E step of six possible substeps, or phases.

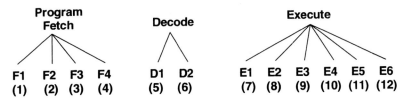

Figure 3-8: Stages of the pipeline.

When the outcome of an instruction is used by the next instruction, an appropriate number of NOPs (no operation or delay) must be added after multiply (one NOP), load (four NOPs/or NOP 4), and branch (five NOPs/or NOP 5) instructions in order to allow the pipeline to operate properly. Therefore, for the above example to run on the C6x processor, appropriate NOPs, as shown in Figure 3-9, should be added after the instructions MPY, LDH, and B.

```
            MVK   .S1    40,A2
loop:       LDH   .D1    *A5++,A0
            LDH   .D1    *A6++,A1
            NOP          4
            MPY   .M1    A0,A1,A3
            NOP
            ADD   .L1    A3,A4,A4
            SUB   .L1    A2,1,A2
      [A2]  B     .S1    loop
            NOP          5
            STH   .D1    A4,*A7
```

Figure 3-9: Pipelined code with NOPs inserted.

Figure 3-10 illustrates an example of a pipeline situation that requires adding an NOP. The plus signs indicate the number of substeps or latencies required for the instruction to be completed. In this example, it is assumed that the addition operation is done before one of its operands is made available from the previous multiply operation, hence the need for adding a NOP after the MPY. Later on, it will be seen that as part of code optimization, NOPs can be reduced or removed leading to an improvement in efficiency.

Prog Fetch	Decode	Execute							Done
F1-F4	D1-D2	E1	E2	E3	E4	E5	E6		
MPY									

MPY is fetched.

Prog Fetch	Decode	Execute							Done
	MPY								
ADD									

MPY is decoded and ADD is fetched.

Prog Fetch	Decode	Execute							Done
		MPY	+						
	ADD								

MPY is executed and ADD is decoded.

Prog Fetch	Decode	Execute							Done
			MPY						
		ADD							

MPY is still being executed while ADD is also executed.

Prog Fetch	Decode	Execute							Done
									► MPY
									► ADD

Both instructions finish at the same time, the result from the MPY is not used in the ADD instruction.

(a)

Prog Fetch	Decode	Execute							Done
F1-F4	D1-D2	E1	E2	E3	E4	E5	E6		
MPY									

MPY is fetched.

Prog Fetch	Decode	Execute							Done
	MPY								
NOP									

MPY is decoded and NOP is fetched.

Prog Fetch	Decode	Execute							Done
		MPY	+						
	NOP								
ADD									

MPY is executed, NOP is decoded and ADD is fetched.

Prog Fetch	Decode	Execute							Done
			MPY						
		NOP							
	ADD								

MPY is still being executed while NOP stalls the pipeline and ADD is decoded.

Prog Fetch	Decode	Execute							Done
									► MPY
		ADD							

MPY completes, ADD is executed while using the result from the MPY.

Prog Fetch	Decode	Execute							Done
									► ADD

ADD completes.

(b)

Figure 3-10: (a) Multiply then add, and (b) need for NOP insertion.

3.3 VelociTI

The C6x architecture is based on the very long instruction word (VLIW) architecture. In such an architecture, several instructions are captured and processed simultaneously. This is referred to as a fetch packet (FP). (See Figure 3-11.)

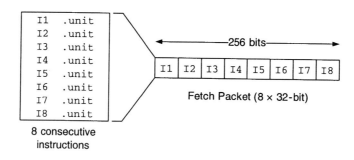

Figure 3-11: C6x fetch packet: C6x fetches eight 32-bit instructions every cycle.

The C6x uses VLIW, allowing eight instructions to be captured simultaneously from on-chip memory onto its 256-bit wide program data bus. The original VLIW architecture has been modified by TI to allow several so-called execute packets (EP) to be included within the same Fetch Packet, as shown in Figure 3-12. An EP constitutes a group of parallel instructions. Parallel instructions are indicated by double pipe symbols (| |), and, as the name implies, they are executed together, or in parallel. Instructions within an EP move together through every stage of the pipeline. This VLIW modification is called VelociTI. Compared with VLIW, VelociTI reduces code size and increases performance when instructions reside off-chip.

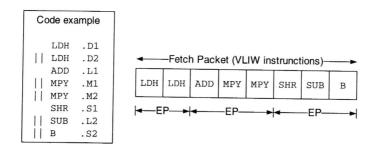

Figure 3-12: A fetch packet containing three execute packets.

3.4 C64x DSP

The C64x is a more recently released DSP core, as part of the C6x family, with higher MIPS power operating at higher clock rates. This core can operate in the range of 300–1000 MHz clock rates, giving a processing power of 2400–8000 MIPS. The clock rate is expected to increase to 1.1 GHz and higher, leading to a processing rate of 8800+ MIPS. The TI website http://www.ti.com/ provides the C64x speed-ups obtained over the C62x for various wireless communication and digital imaging algorithms. Such speedups are achieved due to many enhancements, some of which are mentioned here.

Per CPU data path, the number of registers is increased from 16 to 32, A0–A31 and B0–B31. These registers support packed datatypes, allowing storage and manipulation of four 8-bit or two 16-bit values within a single 32-bit register.

Although the C64x is code compatible with the C62x, (i.e., all the C62x instructions run on the C64x), the C64x can run additional instructions on packed datatypes, boosting parallelism. For example, the new instruction MPYU4 performs four, or quad, 8-bit multiplications, or the instruction MPY2 performs two, or dual, 16-bit multiplications in a single instruction cycle on a .M unit. This packed data processing capability is illustrated in Figure 3-13. Table 3-2 provides a listing of the C64x packed data instructions.

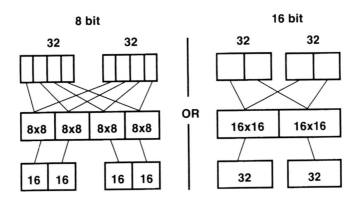

Figure 3-13: C64x packed data processing capability.[†]

Table 3-2: A listing of C64x packed data instructions.†

Operations	Quad 8-bit	Quad 16-bit
Multiply	X	X
Multiply with Saturation		X
Addition/Subtraction	X	X
Addition with Saturation	X	X
Absolute Value		X
Subtract with Absolute Value	X	
Compare	X	X
Shift		X
Data pack/Unpack	X	X
Data pack with Saturation	X	X
Dot Product with optional negate	X	X
Min/Max/Average	X	X
Bit-expansion (Mask generation)	X	X

Additional hardware has been added to each functional unit on the C64x for performing ten special purpose instructions to accelerate key functions encountered in wireless and digital imaging applications. For example, the instruction GMPY4 allows four 8-bit Galois-field multiplications in a single instruction as part of the Reed-Solomon decoding. Table 3-3 provides a list of these special purpose instructions.

Table 3-3: C64x special purpose instructions.†

Instruction	Description	Example Application
BITC4	Bit count	Machine vision
GMPY4	Galois Field MPY	Reed-Solomon support
SHFL	Bit interleaving	Convolution encoder
DEAL	Bit deinterleaving	Cable modem
SWAP4	Byte swap	Mixed Multiprocessor support
XPNDx	Bit expansion	Graphics
MPYHIx, MPYLIx	Extended precision 16x32 MPYs	Audio
AVGx	Quad 8-bit, Dual 16-bit average	Motion compensation
SUBABS4	Quad 8-bit Absolute of differences	Motion estimation
SSHVL, SSHVR	Signed variable shift	GSM

In addition, the functionality of each functional unit on the C64x has been improved leading to a greater orthogonality, or generality, of operations. For example, the .D unit can perform 32-bit logical operation just as the .S and .L units, or the .M unit can perform shift and rotate operations just as the .S unit. The C64x .S unit is capable of performing additional branching instructions, such as branch positive BPOS. Furthermore, the C64x allows multiple units on one side to read the same crosspath source from the other side.

The C64x supports 64-bit loads and stores with a single instruction. There are four 32-bit paths for loading data to the registers. LD1a and LD2a are the load paths for 32 LSBs (least significant bits) on side A and B, respectively, and LD1b and LD2b for 32 MSBs (most significant bits). Similarly, ST1a, ST1b, ST2a, ST2b are the 32-bit store paths for storing data from memory. The C64x also allows nonaligned loads and stores, meaning that loading and storing of words and double words can be done on any byte boundary by using nonaligned load and store instructions. Figure 3-14 provides a comparison between the data paths of the C62x and C64x CPUs.

Finally, similar to the C6211, the C64x contains a 2-level cache, allowing it to make a better use of the CPU speed when interacting with off-chip memory that has a lower speed.

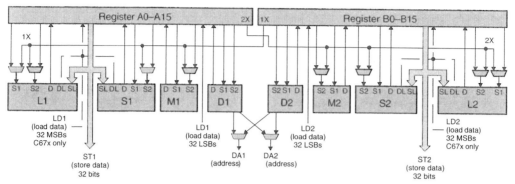

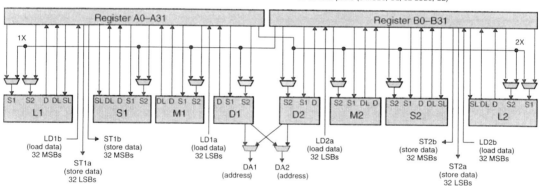

**Figure 3-14: Data paths of C62x/67x and C64x
(S1=source1, S2=source2, D=destination, SL=source long, DL=destination long).**[†]

Bibliography

[1] Texas Instruments, *TMS320C6000 CPU and Instruction Set Reference Guide*, Literature ID# SPRU 189F, 2000.

[2] Texas Instruments, *Technical Training Notes on TMS320C6x*, TI DSP Fest, Houston, 1998.

CHAPTER **4**

Software Tools

Programming most DSP processors can be done either in C or assembly. Although writing programs in C would require less effort, the efficiency achieved is normally less than that of programs written in assembly. Efficiency means having as few instructions or as few instruction cycles as possible by making maximum use of the resources on the chip.

In practice, one starts with C coding to analyze the behavior and functionality of an algorithm. Then, if the required processing rate is not met by using the C compiler optimizer, the time-consuming portions of the C code are identified and converted into assembly, or the entire code is rewritten in assembly. In addition to C and assembly, the C6x allows writing code in linear assembly. Figure 4-1 illustrates the code efficiency versus coding effort for three types of source files on the C6x: C, linear assembly, and hand-optimized assembly. As can be seen, linear assembly provides a good compromise between code efficiency and coding effort.

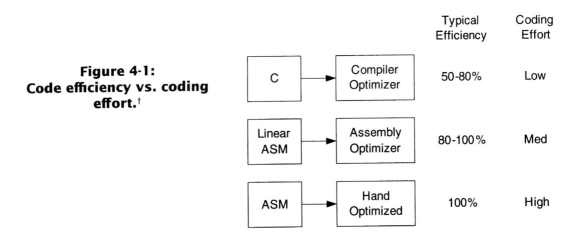

**Figure 4-1:
Code efficiency vs. coding
effort.**[†]

		Typical Efficiency	Coding Effort
C → Compiler Optimizer		50-80%	Low
Linear ASM → Assembly Optimizer		80-100%	Med
ASM → Hand Optimized		100%	High

Figure 4-2 shows the steps involved for going from a source file (.c extension for C, .asm for assembly, and .sa for linear assembly) to an executable file (.out extension). Figure 4-3 lists the .c and .sa versions of the dot-product example to see what they look like. The assembler is used to convert an assembly file into an object file (.obj extension). The assembly optimizer and the compiler are used to convert, respectively, a linear assembly file and a C file into an object file. The linker is used to combine object files, as instructed by the linker command file (.cmd extension), into an executable file. All the assembling, linking, compiling, and debugging steps have been incorporated into an integrated development environment (IDE) called Code Composer Studio (CCS or CCStudio). CCS provides an easy-to-use graphical user environment for building and debugging C and assembly codes on various target DSPs.

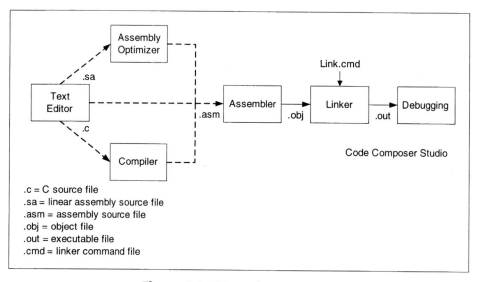

Figure 4-2: C6x software tools.

```
void main()
{
      y = DotP( (int *) a, (int *) x, 40);
}

int DotP(int *m, int *n, short count)
{
      int sum, i;
      sum = 0;

      for(i=0;i<count;i++)
            sum += m[i] * n[i];

      return(sum);
}
```

(a)

```
            .title "dot product"
            .def dotp
            .sect  code

dotp: .proc A4,B4,A6,B6,A8,B3
            .reg  a, ai, b,bi,r,prod,sum,c,ci,i;
            MV    A4,c
            MV    B4,b
            MV    A6,a
            MV    B6,r
            MV    A8,i

loop:       .trip 40
            LDH   *a++, ai
            LDH   *b++,bi
            MPY   ai,bi,prod
            SHR   prod,15,sum
            ADD   ai,sum,ci
            STH   ci, *c++
      [i]   SUB   i,1,i
      [i]   B     loop
            .endproc B3
```

(b)

Figure 4-3: (a) .c (b) .sa version of dot-product example.

4.1 C6x DSK/EVM Target Boards

Upon the availability of either a DSK or an EVM board, an executable file can be run on an actual C6x processor. In the absence of such boards, CCS can be configured to simulate the execution process. As shown in Figure 4-4, the C6713 DSK board is a DSP system which includes a C6713 DSP chip operating at 225 MHz with 4/4/256 Kbytes memory for L1D data cache/L1P program cache/L2 memory, respectively, 8 Mbytes of onboard SDRAM (synchronous dynamic RAM), 512 Kbytes of flash memory, and a 16-bit stereo codec AIC23 with sampling frequency of 8 kHz to 96 kHz. The C6416 DSK board includes a C6416 DSP chip operating at 600 MHz with 16/16/1024 Kbytes memory for L1D data cache/L1P program cache/L2 cache, respectively, 16 Mbytes of onboard SDRAM, 512 Kbytes of flash memory, and AIC23 codec. The C6711 DSK board includes a C6711 DSP chip operating at 150 MHz with 4/4/64 Kbytes of memory for L1D data cache/L1P program cache/L2 cache, 16 Mbytes of onboard SDRAM, 128 Kbytes flash memory, a 16-bit codec AD535 having a fixed sampling frequency of 8 kHz, and a daughter card interface to which a PCM3003 audio daughter card can be connected for changing sampling frequency.

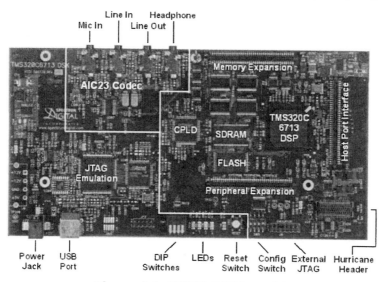

Figure 4-4: C6713 DSK board.[†]

As shown in Figure 4-5(a), the C6701/C6201 EVM board is a DSP system which includes a C6701 (or C6201) chip, external memory, A/D capabilities, and PC host interfacing components. The functional diagram of the EVM board appears in Figure 4-5(b). The board has a 16-bit codec CS4231A whose sampling frequency can be changed from 5.5 kHz to 48 kHz.

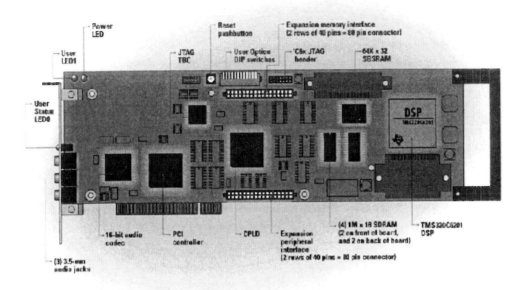

(a)

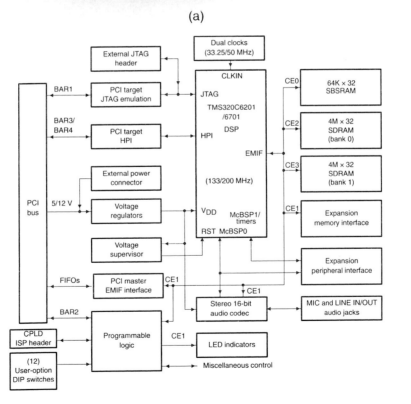

Figure 4-5: (a) C6201/C6701 EVM board, (b) its functional diagram.[†]

The memory residing on the EVM board consists of 32 Mbytes SDRAM running at 100 MHz and 256 Kbytes SBSRAM (synchronous burst static RAM) running at 133 MHz, a faster, but more expensive, memory as compared to SDRAM. A voltage regulator on the board is used to provide 1.8V or 2.5V for the C6x core and 3.3V for its memory and peripherals, and 5V for audio components.

4.2 Assembly File

Similar to other assembly languages, the C6x assembly consists of four fields: label, instruction, operands, and comment. (See Figure 3-4.) The first field is the label field. Labels must start in the first column and must begin with a letter. A label, if present, indicates an assigned name to a specific memory location that contains an instruction or data. Either a mnemonic or a directive constitutes the instruction field. It is optional for the instruction field to include the functional unit which performs that particular instruction. However, to make codes more understandable, the assignment of functional units is recommended. If a functional unit is specified, the data path must be indexed by 1 for the A side and 2 for the B side. A parallel instruction is indicated by a double pipe symbol (| |), and a conditional instruction by a register appearing in brackets in the instruction field. As the name operand implies, the operand field contains arguments of an instruction. Instructions require two or three operands. Except for store instructions, the destination operand must be a register. One of the source operands must be a register, the other a register or a constant. After the operand field, there is an optional comment field that, if stated, should begin with a semicolon (;).

4.2.1 Directives

Directives are used to indicate assembly code sections and to declare data structures. It should be noted that assembly statements appearing as directives do not produce any executable code. They merely control the assembling process by the assembler. Some of the widely used assembler directives are:

.sect "name" directive, which defines a section of code or data named "name".

.int, .long, or .word directive, which reserves 32 bits of memory initialized to a value.

.short or .half directive, which reserves 16 bits of memory initialized to a value.

.byte directive, which reserves 8 bits of memory initialized to a value.

Note that in the TI common object file format (COFF), the directives .text, .data, .bss are used to indicate code, initialized constant data, and uninitialized variables, respectively. Other directives often used include .set directive, for assigning a value to a symbol, .global or .def directive, to declare a symbol or module as global so that it may be recognized externally by other modules, and .end directive, to signal the termination of assembly code. The directive .global acts as a .def directive for defined symbols and as a .ref directive for undefined symbols.

At this point, it should be mentioned that the C compiler creates various sections indicated by the directives .text, .switch, .const, .cinit, .bss, .far, .stack, .sysmem, .cio. Figure 4-6 lists some common compiler sections. For a complete listing of directives, refer to the *TI TMS320C6x Assembly Language Tools User's Guide* [1].

Section Name	Description
.text	Code
.switch	Tables for switch instructions
.const	Global and static string literals
.cinit	Initial values for global/static vars
.bss	Global and static variables
.far	Global and statics declared far
.stack	Stack (local variables)
.sysmem	Memory for malloc fcns (heap)
.cio	Buffers for stdio functions

Figure 4-6: Common compiler sections.

4.3 Memory Management

The external memory used by a DSP processor can be either static or dynamic. Static memory (SRAM) is faster than dynamic memory (DRAM), but it is more expensive, since it takes more space on silicon. DRAMs also need to be refreshed periodically. A good compromise between cost and performance is achieved by using SDRAM (Synchronous DRAM). Synchronous memory requires clocking, as compared to asynchronous memory, which does not.

Given that the address bus is 32 bits wide, the total memory space consists of 2^{32} = 4 Gbytes. On the EVM, this space is divided, according to a memory map, into the internal program memory (PMEM), internal data memory (DMEM), internal peripherals, and external memory spaces named CE0, CE1, CE2, and CE3. There are two memory map configurations: memory map 0 and memory map 1. Figures 4-7(a) and 4-7(b) illustrate these two memory maps. On the DSK, there is no separation between internal program and data memory. For the lab exercises in this book, the EVM board is configured based on its memory map 1 as shown in Figure 4-7(c), and the DSK board based on its memory map 1 as shown in Figure 4-7(d).

Address	Memory Map 0	Block Size (Bytes)
0000 0000	External-Memory Space CE0	16M
0100 0000	External-Memory Space CE1	4M
0140 0000	Internal Program RAM	64K
0141 0000	Reserved	4M
0180 0000	Internal Peripheral Space	4M
01C0 0000	Reserved	4M
0200 0000	External Memory Space CE2	16M
0300 0000	External Memory Space CE3	16M
0400 0000	Reserved	1984M
8000 0000	Internal Data RAM	64K
8001 0000	Reserved	4M
8040 0000	Reserved	2044M
1 0000 000		

(a)

Address	Memory Map 0	Block Size (Bytes)
0000 0000	Internal Program RAM	64K
0001 0000	Reserved	4M
0040 0000	External-Memory Space CE0	16M
0140 0000	External Memory Space CE1	4M
0180 0000	Same as Memory Map 0	
1 0000 000		

(b)

Address	Memory Map 1	Block Size (Bytes)	
0000 0000	Internal Program RAM	64K	4M
0001 0000	Reserved		
0040 0000	SBSRAM	256K	16M
0044 0000	Unused	Unused	
0140 0000	Asynchronous Expansion Memory	3M	
0170 0000	PCI add-on registers	64	64K
0170 0040	Unavailable		
0171 0000	PCI FIFO	4	64K
0171 0004	Unavailable		16M
0172 0000	Audio Codec Registers	16	64K
0172 0010	Unavailable		
0173 0000	Reserved	320K	
0178 0000	DSP control/status registers	32	64K
0178 0020	Unavailable		
0179 0000	Reserved	448K	
0180 0000	Internal Peripheral Space	4M	
01C0 0000	Reserved	4M	
0200 0000	SDRAM (Bank 0)	4M	
0240 0000	Reserved	12M	
0300 0000	SDRAM (Bank 1)	4M	
0340 0000	Reserved	12M	
0400 0000	Reserved	1984M	
8000 0000	Internal Data RAM	64K	4M
8001 0000	Reserved		
8040 0000	Reserved	2044M	
10000 0000			

(c)

Address	Memory Map 1	Block Size (Bytes)
0000 0000	Internal RAM (L2)	64K
0001 0000	Reserved	24M
0180 0000	EMIF control regs	32
0184 0000	Cache Configuration regs	4
0184 4000	L2 base addr & count regs	32
0184 4020	L1 base addr & count regs	32
0184 5000	L2 flush & clean regs	32
0184 8200	CE0 mem attribute regs	16
0184 8240	CE1 mem attribute regs	16
0184 8280	CE2 mem attribute regs	16
0184 82C0	CE3 mem attribute regs	16
0188 0000	HPI control regs	4
018C 0000	McBSP0 regs	40
0190 0000	McBSP1 regs	40
0194 0000	Timer0 regs	12
0198 0000	Timer1 regs	12
019C 0000	Interrupt selector regs	12
01A0 0000	EDMA parameter RAM	2M
01A0 FFE0	EDMA control regs	32
0200 0000	QDMA regs	20
0200 0020	QDMA pseudo-regs	20
3000 0000	McBSP0 data	64M
3400 0000	McBSP1 data	64M
8000 0000	CE0, SDRAM	16M
9000 0000	CE1, 8-bit ROM	128K
9008 0000	CE1, 8-bit I/O port	4
A000 0000	CE2-Daughtercard	256M
B000 0000	CE3-Daughtercard	256M
10000 0000		

(d)

Figure 4-7: (a) C6x memory map 0, (b) map 1, (c) EVM map 1, and (d) DSK map 1.[†]

The external memory ranges CE0, CE1, CE2, and CE3 support synchronous (SB-SRAM, SDRAM) or asynchronous (SRAM, ROM, and so forth) memory, accessible as bytes (8 bits), halfwords (16 bits), or words (32 bits). The on-chip peripherals and control registers are mapped into the memory space. A listing of the memory-mapped registers is provided in Appendix A (Quick Reference Guide).

The internal data memory is organized into memory banks so that two loads or stores can be done simultaneously. As long as data are accessed from different banks, no conflict occurs. However, if data are accessed from the same bank in one instruction, a memory conflict occurs and the CPU is stalled by one cycle.

If a program fits into the on-chip or internal memory, it should be run from there to avoid delays associated with accessing off-chip or external memory. If a program is too big to be fitted into the internal memory, most of its time-consuming portions should be placed into the internal memory for efficient execution. For repetitive codes, it is recommended that the internal memory is configured as cache memory. This allows accessing external memory as seldom as possible and hence avoiding delays associated with such accesses.

4.3.1 Linking

Linking places code, constant, and variable sections into appropriate locations in memory as specified in the .cmd linker command file. Also, it combines several .obj object files into the final executable .out output file. A typical command file corresponding to the DSK memory map 1 is shown below in Figure 4-8.

The first part, MEMORY, provides a description of the type of physical memory, its origin and its length. The second part, SECTIONS, specifies the assignment of various code sections to the available physical memory.

```
MEMORY
{
        VECS:   o=00000000h   l=00000200h     /* interrupt vectors */
        PMEM:·  o=00000200h   l=0000FE00h     /* Internal RAM (L2) mem */
        BMEM:   o=80000000h   l=01000000h     /* CE0, SDRAM, 16 Mbytes */
}

SECTIONS
{
       .intvecs     >  0h
       .text        >  PMEM
       .far         >  PMEM
       .stack       >  PMEM
       .bss         >  PMEM
       .cinit       >  PMEM
       .pinit       >  PMEM
       .cio         >  PMEM
       .const       >  PMEM
       .data        >  PMEM
       .switch      >  PMEM
       .sysmem      >  PMEM
}
```

Figure 4-8: A typical linker command file.

4.4 Compiler Utility

The build feature of CCS can be used to perform the entire process of compiling, assembling, and linking in one step via the activation of the utility cl6x and stating the right options for it. The following command shows how this utility is used within CCS for building the source files *file1.c*, *file2.asm*, and *file3.sa*:

```
cl6x -gs file1.c file2.asm file3.sa  -z  -o file.out  -m file.map  -l rts6700.lib
```

The option -g adds debugger specific information to the object file for debugging purposes. The option -s provides an interlisting of C and assembly. For *file1.c*, the C compiler, for *file2.asm* the assembler, and for *file3.sa*, the assembly optimizer (linear assembler) are invoked. The option -z invokes the linker, placing the executable code in *file.out* if the -o option is used. Otherwise, the default file *a.out* is created. The option -m provides a map file (*file.map*), which includes a listing of all the addresses of sections, symbols and labels. The option -l specifies the run-time support library *rts6700.lib* for linking files on the C6713 processor. Table 4-1 lists some frequently used options. Refer to the *TI Optimizing C Compiler* manual [2] for a complete list of available options.

The compiler allows four levels of optimizations to be invoked by using -o0, -o1, -o2, -o3. Debugging and full-scale optimization cannot be done together, since they are in conflict; that is, in debugging, information is added to enhance the debugging process, while in optimizing, information is minimized or removed to enhance code efficiency. In essence, the optimizer changes the flow of C code, making program debugging very difficult.

As shown in Figure 4-9, a good programming approach would be first to verify that the code is properly functioning by using the compiler with no optimization (-gs option). Then, use full optimization to generate an efficient code (-o3 option). It is recommended that an intermediary step be taken in which some optimization is done without interfering with source level debugging (-go option). This intermediary step can reverify code functionality before performing full optimization. It should be pointed out that full optimization may change memory locations outside the scope of the C code. Such memory locations must be declared as 'volatile' to prevent compiling errors.

Table 4-1: Common compile options.[†]

Options	Description	Tool
-mv6700	Generate 'C67x code ('C62x is default)	Comp/Asm
-g	Enables src-level symbolic debugging	Comp/Asm
-mg	Enables minimum debug to allow profiling	Compiler
-s	Interlist C statements into assembly listing	Compiler
-o	Invoke optimizer (-o0, -o1, -o2/ -o, -o3)	Compiler
-pm	Combine all C source files before compile	Compiler
-mt	No aliasing used	Compiler
-ms	Minimize code size (-ms0/ -ms, -ms1, -ms2)	Compiler
-z	Invokes linker	Linker
-o	Output file name	Linker
-m	Map file name	Linker
-c	Auto -Init C variables (-cs turns off autoinit)	Linker
-l	Link -in libraries (small -L)	Linker

```
1.  Compile without optimization.
    (Get the code functioning!)
    cl6x -g -s file.c -z

2.  Compile with some optimization.
    (Verify code functionality, again)
    cl6x -g -o file.c -z

3.  Compile with all optimizations.
    (Generate efficient code)
    cl6x -o3 -pm file.c -z
```

Figure 4-9: Programming approach.

As a step to further optimize C codes, it is recommended that *intrinsics* be used wherever possible. Intrinsics are functions similar to math functions as part of the runtime support library. Intrinsics allow the C compiler to directly access the hardware while preserving the C environment. As an example, instead of using the multiply operator * in C, the intrinsic _mpy() can be used to tell the compiler to use the C6x instruction MPY. Figure 4-10 shows the intrinsic version of the dot-product C code. A list of the C6x intrinsics is provided in Appendix A (Quick Reference Quide).

```c
short DotP(int *m, int *n, short count)
{
        short i, productl, producth, suml = 0, sumh = 0;

        for(i=0; i<count; i++)
        {
                productl = _mpy(m[i],n[i]);        // _mpy intrinsic
                producth = _mpyh(m[i],n[i]);       // _mpyh intrinsic
                suml += productl;
                sumh += producth;
        }
        suml += sumh;
        return(suml);
}
```

Figure 4-10: Intrinsic version of dot-product C code.

4.5 Code Initialization

All programs start by going through a reset initialization code. Figure 4-11 illustrates both the C and assembly version of a typical reset initialization code. This initialization is for the purpose of starting at a previously defined initial location. Upon power up, the system always goes to the reset location in memory, which normally includes a branch instruction to the beginning of the code to be executed. The reset code shown in Figure 4-11 takes the program counter to a globally defined location in memory named `init` or `_c_int00`.

"ASM"

```
vectors.asm

        .ref    init
        .sect   "vectors"
rst     MVK     .s2    init,B0
        MVKH    .s2    init,B0
        B       .s2    B0
        NOP
        NOP
        NOP
        NOP
        NOP
```

"C"

```
cvectors.asm

        .global _c_int00
        .sect   "vectors"
rst     B       _c_int00
        NOP     ;additional NOP's
        NOP     ;to create a
        NOP     ;fetch packet
        NOP
        NOP
        NOP
        NOP
```

Figure 4-11: Reset code.

As indicated in Figure 4-12, when writing in assembly, an initialization code is needed to create initialized data and variables, and to copy initialized data into corresponding variables. Initialized values are specified by using `.byte`, `.short`, or `.int` directives. Uninitialized variables are specified by using `.usect` directive. The first, second, and third arguments of this directive denote section name, size in bytes, and data alignment in bytes, respectively. Before calling the main function or subroutine, another initialization code portion is usually needed to set up registers and pointers, and to move data to appropriate places in memory.

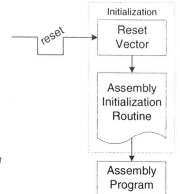

Figure 4-12: Assembly initialization.[†]

Figure 4-13 provides the initialization code for the dot-product example in which initialized data values appear for three initialized data arrays labeled `table_a`, `table_x`, and `table_y`. In addition, three variable sections called a, x, and y are declared. The second part of the initialization code copies the initialized data into the corresponding variables. The setup code for calling the dot-product routine is also shown in this figure.

```
        .def    init
        .ref    dotp

;Data initialization
;Initialize tables

        .sect   "init_tables"

table_a .short  40,39,38,37,36,35,34,33,32,31,30,29,28,27  ;Initialize table_a array with values
        .short  26,25,24,23,22,21,20,19,18,17
        .short  16,15,14,13,12,14,10,9,8,7,6,5,4,3,2,1
table_x .short  1,2,3,4,5,6,7,8,9,10,11,12,13,14,15        ;Initialize table_x array with values
        .short  16,17,18,19,20,21,22,23,24,25,26,27,28,29
        .short  30,31,32,33,34,35,36,37,38,39,40
table_y .short  0                                          ;table_y = 0

;Variable declaration
a       .usect  "var", 80, 2                                ;define variables
x       .usect  "var", 80, 2
y       .usect  "var", 2, 2

;Initialization to copy data into variables
        .sect   "init_code"
init    mvk     .s1     table_a, A0     ;move address of table_a to register A0
        mvkh    .s1     table_a, A0
        mvk     .s2     a,B0            ;move address of a to register B0
        mvkh    .s2     a,B0
        mvk     .s2     40,B1           ;create a counter in register B1, B1=40
loop_a  ldh     .d1     *A0++,A1        ;load an element from the address pointed by A0 into A1
        sub     .l2     B1,1,B1         ;decrement counter
        nop             3
        sth     .d2     A1,*B0++        ;store the element to address pointed by B0
   [B1] b       .s2     loop_a          ;branch back to loop_a
        nop             5               ;required latency
init_x  mvk     .s1     table_x, A0     ;move address of table_x into register A0
        mvkh    .s1     table_x, A0
        mvk     .s2     x, B0           ;move address of x into register A0
        mvkh    .s2     x, B0
        mvk     .s2     40, B1          ;create a counter
loop_x  ldh     .d1     *A0++,A1        ;load an element from the address pointed by A0 into A1
        sub     .l2     B1,1,B1         ;decrement counter
        nop             3
        sth     .d2     A1,*B0++        ;store element to address pointed by B0
   [B1] b       .s2     loop_x          ;branch back to loop_x
        nop             5
init_y  mvk     .s1     table_y, A0     ;repeat above procedure for table_y
        mvkh    .s1     table_y, A0
        mvk     .s2     y, B0
        mvkh    .s2     y, B0
        ldh     .d1     *A0, A1
        nop             4
        sth     .d2     A1, *B0
```

(a)

```
;Setup for calling dotp

start   mvk     .s1     a,A4            ;move a into register A4
        mvkh    .s1     a,A4
        mvk     .s2     x,B4            ;move x into register B4
        mvkh    .s2     x,B4
        mvk     .s1     40,A6           ;create a counter in A6, A6=40
        b       .s1     dotp            ;branch to routine dotp
        mvk     .s2     return, B3      ;store return address in B3
        mvkh    .s2     return, B3
        nop             3

;return from dotp here
return  mvk     .s1     y, A0           ;move y into register A0
        mvkh    .s1     y, A0
        sth     .d1     A4, *A0         ;store the result of dotp (returned in A4) to y

;infinite loop
end     b       .s1     end             ;infinite loop
        nop             5
```

(b)

```
;dotp
        .def    dotp

;A4 = &a, B4 = &x, A6 = 40 (iteration count) , B3 = return address

dotp    mv              A6,B0           ;move A6 to B0 (third argument passed from calling function)
        zero            A2              ;zero the sum register A2

loop    ldh     .d1     *A4++,A5        ;load an element from the location pointed by A4 into A5
        ldh     .d2     *B4++,B5        ;load an element from the location pointed by B4 into B5
        nop             4
        mpy     .m1x    A5,B5,A5        ;A5=B5*A5
        nop
        add     .l1     A2,A5,A2        ;A2 += A5
   [B0] sub     .l2     B0,1,B0         ;decrement counter B0
   [B0] b       .s1     loop            ;branch back to loop
        nop             5

        mv              A2,A4           ;move result in A2 to return register A4
        b       .s2     B3              ;branch back to calling address stored in B3
        nop             5
```

(c)

**Figure 4-13: (a) Initialization code for dot-product example,
(b) setup code for calling dot product routine, and (c) dot product routine.** [†]

As far as C coding is concerned, the C compiler uses *boot.c* in the run-time support library to perform the initialization before calling main(). The –c option activates *boot.c* to autoinitialize variables. This is illustrated in Figure 4-14.

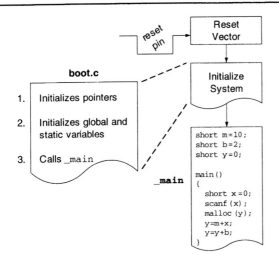

Figure 4-14: C initialization.[†]

4.5.1 Data Alignment

The C6x allows byte, half-word, or word addressing. Consider a word-format representation of memory as shown in Figure 4-15. There are four byte boundaries, two half-word (or short) boundaries, and one word boundary per word. The C6x always accesses data on these boundaries depending on the datatype specified; that is, it always accesses aligned data. When specifying an uninitialized variable section .usect, it is required to specify the alignment as well as the total number of bytes. The examples appearing in Figure 4-16 show data alignment for both constants and variables.

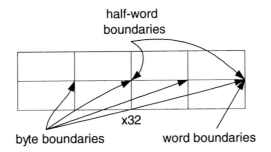

Figure 4-15: Data boundaries.

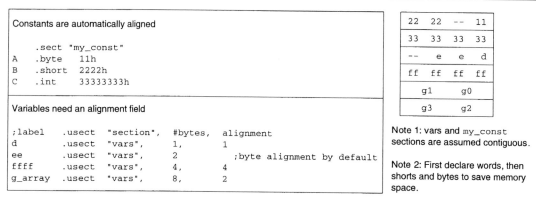

Figure 4-16: Constant and variable alignment examples.[†]

Data in memory can be arranged either in little- or big-endian format. Little-endian (le) means that the least significant byte is stored first. Figure 4-17(a) shows storing .int 40302010h in little-endian format for byte, half-word, and word access addressing. In big-endian (be) format, shown in Figure 4-17(b), the most significant byte is stored first. Little-endian is the format normally used in most applications. Additional data alignment examples are shown in Figure 4-17(c), based on the little-endian data format appearing in Figure 4-17(a).

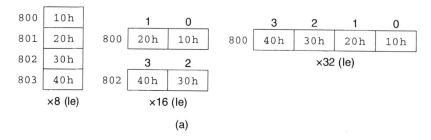

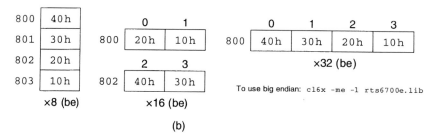

Example code		A0	A1	Pointer aligned on datatype boundary?
LDH	*A0,A1	800h	0000 2010h	Yes
LDB	*A0++,A1	800h	0000 0010h	Yes
LDH	*A0,A1	801h	0000 2010h	No
LDW	*A0,A1	801h	4030 2010h	No
LDW	*++A0,A1	805h	8070 6050h	No

Figure 4-17: (a) Little endian, (b) big endian, and (c) more data alignment examples.[†]

Bibliography

[1] Texas Instruments, *TMS320C6000 Assembly Language Tools User's Guide*, Literature ID# SPRU 186M, 2003.

[2] Texas Instruments, *TMS320C6000 Optimizing Compiler User's Guide*, Literature ID# SPRU 187K, 2002.

Lab 1: Getting Familiar with Code Composer Studio

Code Composer Studio™ (CCStudio or CCS) is a useful integrated development environment (IDE) that provides an easy-to-use software tool to build and debug programs. In addition, it allows real-time analysis of application programs. Figure 4-18 shows the phases associated with the CCS code development process. During its set-up, CCS can be configured for different target DSP boards (for example, C6711 DSK, C6416 DSK, C6701 EVM, C6xxx Simulator). The version used throughout the book is based on CCS 2.2, the latest version at the time of this writing.

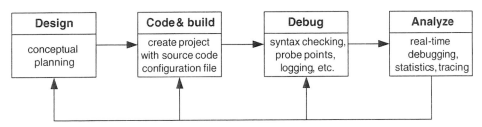

Figure 4-18: CCS code development process.[†]

CCS provides a file management environment for building application programs. It includes an integrated editor for editing C and assembly files. For debugging purposes, it provides breakpoints, data monitoring and graphing capabilities, profiler for benchmarking, and probe points to stream data to and from the target DSP. This tutorial lab introduces the basic features of CCS that are needed in order to build and debug an application program. To become familiar with all of its features, one should go through the *TI CCS Tutorial* and *TI CCS User's Guide* manuals [1-2] . The real-time analysis and scheduling features of CCS will be covered later, in Labs 7 and 8.

This lab demonstrates how a simple multifile algorithm can be compiled, assembled and linked by using CCS. First, several data values are consecutively written to memory. Then, a pointer is assigned to the beginning of the data so that they can be treated as an array. Finally, simple functions are added in both C and assembly to illustrate how function calling works. This method of placing data in memory is simple to use and can be used in applications in which constants need to be in memory, such as filter coefficients and FFT twiddle factors. Issues related to debugging and benchmarking are also covered in this lab.

The lab programs can be downloaded from the accompanying CD-ROM for different DSP target boards. Six versions of the lab programs appear on the CD. These versions correspond to (a).C6713 DSK, (b) C6416 DSK, (c) C6711 DSK together with a daughter card, (d) C6711 DSK without a daughter card, (e) C6701/C6201 EVM, and (f) simulator.

L1.1 Creating Projects

Let us consider all the files required to create an executable file; that is, `.c` (c), `.asm` (assembly), `.sa` (linear assembly) source files, a `.cmd` linker command file, a `.h` header file, and appropriate `.lib` library files. The CCS code development process begins with the creation of a so-called Project to easily integrate and manage all these required files for generating and running an executable file. The **Project View** panel on the left-hand side of the CCS window provides an easy mechanism for doing so. In this panel, a project file (`.prj` extension) can be created or opened to contain not only the source and library files but also the compiler, assembler, and linker options for generating an executable file. As a result, one need not type command lines for compilation, assembling and linking, as was the case with the original software development tools.

To create a project, choose the menu item **Project** → **New** from the CCS menu bar. This brings up the dialog box **Project Creation**, as shown in Figure 4-19. In the dialog box, navigate to the working folder, throughout the book assumed to be `C:\ti\`

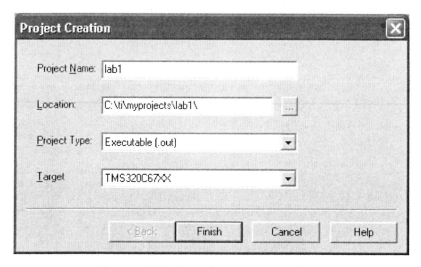

Figure 4-19: Creating a new project.

myprojects, and type a project name in the field **Project Name**. Then, click the button **Finish** for CCS to create a project file named *lab1.prj*. All the files necessary to build an application should be added to the project.

CCS provides an integrated editor which allows the creation of source files. Some of the features of the editor are color syntax highlighting, C blocks marking in parentheses and braces, parenthesis/brace matching, control indentions, and find/re-place/search capabilities. It is also possible to add files to the project from Windows Explorer using the drag-and-drop approach. An editor window is brought up by choosing the menu item **File** → **New** → **Source File**. For this lab, let's type the following assembly code into the editor window:

```
.sect       ".mydata"
.short      0
.short      7
.short      10
.short      7
.short      0
.short      -7
.short      -10
.short      -7
.short      0
.short      7
```

This code consists of the declaration of 10 values using .short directives. Note that any of the data to memory allocation directives can be used to do the same. Assign a section named .mydata to the values by using a .sect directive. Save the created source file by choosing the menu item **File** → **Save**. This brings up the dialog box **Save As**, as shown in Figure 4-20. In the dialog box, go to the field **Save as type** and select **Assembly Source Files (*.asm)** from the pull-down list. Then, go to the field **File name**, and type *initmem.asm*. Finally, click **Save** so that the code can be saved into an assembly source file named *initmem.asm*.

In addition to source files, a linker command file must be specified to create an executable file and to conform to memory specifics of the target DSP on which the executable file is going to run. A linker command file can be created by choosing **File** → **New** → **Source File**. For this lab, let's type the command file shown in Figure 4-21. This file can also be downloaded from the accompanying CD-ROM. This linker command file is configured based on the DSK memory map. Since our intention is to place the array of values defined in *initmem.asm* into the memory, a space that will not be overwritten by the compiler should be selected. The external memory space CE0 can be used for this purpose. Let us assemble the data at the memory address

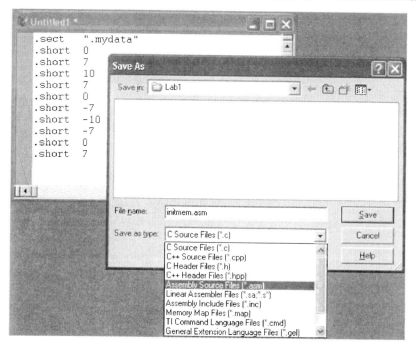

Figure 4-20: Creating a source file.

0x80000000 (0x denotes hex) located at the beginning of CE0. This is done by assigning the section named .mydata to MYDATA via adding .mydata > MYDA-TA to the SECTIONS part of the linker command file, as shown in Figure 4-21. Save the editor window into a linker command file by choosing **File** → **Save** or by pressing **Ctrl + S**. This brings up the dialog box **Save As**. Go to the field **Save as type** and select **TI Command Language Files (*.cmd)** from the pull-down list. Then, type lab1.cmd in the field **File name** and click **Save**.

Now that the source file *initmem.asm* and the linker command file *lab1.cmd* are created, they should be added to the project for assembling and linking. To do this, choose the menu item **Project** → **Add Files to Project**. This brings up the dialog box **Add Files to Project**. In the dialog box, select *initmem.asm* and click the button **Open**. This adds *initmem.asm* to the project. In order to add *lab1.cmd*, choose **Project** → **Add Files to Project**. Then, in the dialog box **Add Files to Project**, set **Files of type** to **Linker Command File (*.cmd)**, so that *lab1.cmd* appears in the dialog box. Now, select *lab1.cmd* and click the button **Open**. In addition to initmem.asm and *lab1.cmd* files, the run-time support library file should be added to the project. To do so, choose **Project** →

```
MEMORY
{
    IRAM:         o = 00000000h     l = 00100000h
    MYDATA:       o = 80000000h     l = 00000100h
    CE0:          o = 80000100h     l = 000FFF00h
}

SECTIONS
{
    .cinit        >      IRAM
    .text         >      IRAM
    .stack        >      IRAM
    .bss          >      IRAM
    .const        >      IRAM
    .data         >      IRAM
    .far          >      IRAM
    .switch       >      IRAM
    .sysmem       >      IRAM
    .tables       >      IRAM
    .cio          >      IRAM
    .mydata       >      MYDATA
}
```

Figure 4-21: Linker command file for Lab 1.

Add Files to Project, go to the compiler library folder, here assumed to be the default option C:\ti\c6000\cgtools\lib, select **Object and Library Files (*.o*,*.l*)** in the box **Files of type**, then select *rts6700.lib* and click **Open**. If running on the fixed-point DSP TMS320C6211, select *rts6200.lib* instead. For debugging purposes, let us use the following empty shell C program. Create a C source file main.c, enter the following lines and add *main.c* to the project in the same way as just described.

```c
#include <stdio.h>

void main()
{
    printf("BEGIN\n");

    printf("END\n");
}
```

After adding all the source files, the command file and the library file to the project, it is time to either build the project or to create an executable file for the target DSP. To do this, choose the **Project** → **Build** menu item. CCS compiles, assembles, and links all of the files in the project. Messages about this process are shown in a panel at the bottom of the CCS window. When the building process is completed without any errors, the executable *lab1.out* file is generated. It is also possible to do incremental builds – that is rebuilding only those files changed since last build, by choosing the menu item **Project** → **Rebuild**. The CCS window provides shortcut buttons for frequently used menu options, such as **build** ⊞ and **rebuild all** ⊞.

Although CCS provides default build options, these options can be changed by choosing **Project** → **Build Options**. For instance, to change the executable filename to *test.out*, choose **Project** → **Build Options**, click the **Linker** tab of the **Build Options** window, and type `test.out` in the field **Output Filename (-o)**, as shown in Figure 4-22a. Notice that the linker command file will include *test.out* as you click on this window.

All the compiler, assembler, and linker options are set through the menu item **Project** → **Build Options**. Among many compiler options shown in 4-22b, pay particular attention to the optimization level options. There are four levels of optimization (0, 1, 2, 3), which control the type and degree of optimization. Level 0 optimization option performs control-flow-graph simplification, allocates variables to registers, eliminates unused code, and simplifies expressions and statements. Level 1 optimization performs all Level 0 optimizations, removes unused assignments, and eliminates local common expressions. Level 2 optimization performs all Level 1 optimizations, plus software pipelining, loop optimizations, and loop unrolling. It also eliminates global common subexpressions and unused assignments. Finally, Level 3 optimization performs all Level 2 optimizations, removes all functions that are never called, and simplifies functions with return values that are never used. It also inlines calls to small functions and reorders function declarations.

Note that in some cases, debugging is not possible due to optimization. Thus, it is recommended to first debug your program to make sure that it is logically correct before performing any optimization. Another important compiler option is the **Target Version** option. When the application is for the floating-point target DSP TMS320C6711/6713 DSK, go to the **Target Version** field and select **671x (–mV 6710)** from the pull-down list. For the fixed-point target DSP TMS320C6416 DSK, select **C64xx (–mv 6400)**. For the EVM target TMS320C6701, select **C670x**.

Build Options for Lab1a.pjt

General | Compiler | Linker | Link Order

-q -c -o".\Debug\Lab1.out" -x

Category:
Basic
Advanced

Basic
☑ Suppress Banner (-q)
☑ Exhaustively Read Libraries (-x)

Output Module:

Output Filename (-o): .\Debug\test.out

Map Filename (-m):

Autoinit Model: Run-time Autoinitialization (-c) ▼

Heap Size (-heap):

Stack Size (-stack):

Fill Value (-f):

Code Entry Point (-e):

Library Search Path (-i):

Include Libraries (-l):

OK   Cancel   Help

(a)

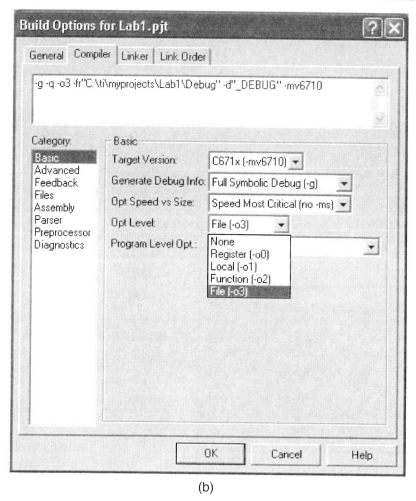

(b)

Figure 4-22: (a) Build options, and (b) compiler options.

One common mistake in writing *initmem.asm* is to type directives .sect and .short in the first column. Because only labels can start in the first column, this will result in an assembler error. When a message stating a compilation error appears, click **Stop Build** and scroll up in the **Build** area to see the syntax error message. Double-click on the red text that describes the location of the syntax error. Notice that the *initmem.asm* file opens, and your cursor appears on the line that has caused the error, see Figure 4-23. After correcting the syntax error, the file should be saved and the project rebuilt.

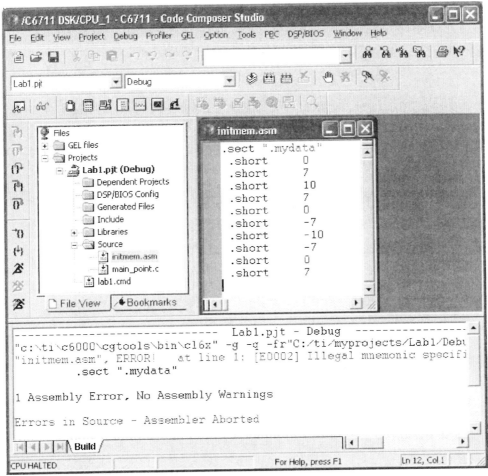

Figure 4-23: Build Error.

L1.2 Debugging Tools

Once the build process is completed without any errors, the program can be loaded and executed on the target DSP. To load the program, choose **File → Load Program**, select the program *lab1.out* just rebuilt, and click **Open**. To run the program, choose the menu item **Debug → Run**. You should be able to see BEGIN and END appearing in the **Stdout** window.

Now, let us verify if the array of values is assembled into the specified memory location. CCS allows one to view the content of memory at a specific location. To view the content of memory at 0x80000000, select **View** → **Memory** from the menu. The dialog box **Memory Window Options** will appear. This dialog box allows one to specify various attributes of the **Memory** window. Go to the **Address** field and enter 0x80000000. Then, select **16-bit Signed Int** from the pull-down list in the **Format** field and click **OK**. A **Memory** window appears as shown in Figure 4-24. The contents of CPU, peripheral, DMA, and serial port registers can also be viewed by selecting **View** → **Registers** → **Core Registers**, for example.

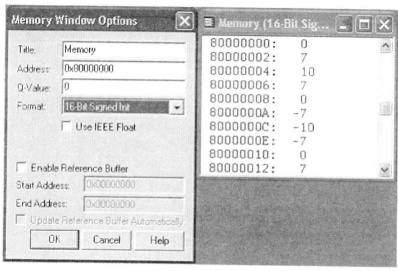

Figure 4-24: Memory Window Options dialog box and Memory window.

There is another way to load data from a PC file to the DSP memory. CCS provides a probe point capability, so that a stream of data can be moved from the PC host file to the DSP or vice versa. In order to use this capability, a Probe Point should be set within the program by placing the mouse cursor at the line where a stream of data needs to be transferred and clicking the button **Probe Point** . Then, choose **File** → **File I/O** to invoke the dialog box **File I/O**. Click the button **Add File** and select the data file to load. Now the file should be connected to the probe point by clicking the button **Add Probe Point**. In the **Probe Point** field, select the probe point to make it active, then connect the probe point to the PC file through **File In:...** in the field **Connect To**. Click the button **Replace** and then the button **OK**. Finally, enter the memory location in the **Address** field and the number of data in the **Length** field. Note that a global variable name can be used in the **Address** field. The probe point capability is frequently used to simulate the execution of an application program in the absence of any live signals. A valid PC file should have the correct file header and extension. The file header should conform to the following format:

```
MagicNumber Format StartingAddress PageNum Length
```

`MagicNumber` is fixed at `1651`. `Format` indicates the format of samples in the file: 1 for hexadecimal, 2 for integer, 3 for long, and 4 for float. `StartingAddress` and `PageNum` are determined by CCS when a stream of data is saved into a PC file. `Length` indicates the number of samples in the memory. A valid data file should have the extension `.dat`.

A graphical display of data often provides better feedback about the behavior of a program. CCS provides a signal analysis interface to monitor a signal or data. Let us display the array of values at `0x80000000` as a signal or a time graph. To do so, select **View** → **Graph** → **Time/Frequency** to view the **Graph Property Dialog** box. Field names appear in the left column. Go to the **Start Address** field, click it and type `0x80000000`. Then, go to the **Acquisition Buffer Size** field, click it and enter `10`. Finally, click on **DSP Data Type**, select **16-bit signed integer** from the pull-down list, and click **OK**. A graph window appears with the properties selected. This is illustrated in Figure 4-25. You can change any of these parameters from the graph window by right-clicking the mouse, selecting **Properties**, and adjusting the properties as needed. The properties can be updated at any time during the debugging process.

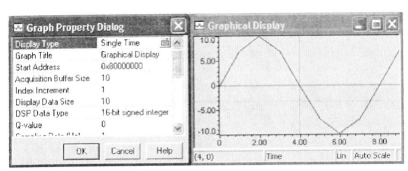

Figure 4-25: Graph Property Dialog box and Graphical Display window.

To assign a pointer to the beginning of the assembled memory space, the memory address can be typed in directly to a pointer. It is necessary to typecast the pointer to short since the values are of that type. The following code can be used to assign a pointer to the beginning of the values and loop through them to print each onto the **Stdout** window:

```
#include <stdio.h>

void main()
{
     int i;
     short *point;
     point = (short *) 0x80000000;

     printf("BEGIN\n");

     for(i=0;i<10;i++)
     { printf("[%d] %d\n",i, point[i]);  }

     printf("END\n");
}
```

Instead of creating a new source file, we can modify the existing *main.c* by double-clicking on the *main.c* file in the **Project View** panel, as shown in Figure 4-26. This action will bring up the *main.c* source file in the right-half part of the CCS window. Then, enter the code and rebuild it. Before running the executable file, make sure you reload the file *lab1.out*. By running this file, you should be able to see the values in the **Stdout** window. Double-clicking in the **Project View** panel provides an easy way to bring up any source or command file for reviewing or modifying purposes.

When developing and testing programs, one often needs to check the value of a variable during program execution. This can be achieved by using breakpoints and watch windows. To view the values of the pointer in *main.c* before and after the pointer assignment, choose **File** → **Reload Program** to reload the program. Then, double-click on *main.c* in the **Project View** panel. You may wish to make the window larger so that you can see more of the file in one place. Next, put your cursor on the line that says point = (short *) 0x80000000 and press F9 to set a breakpoint. To open a watch window, choose **View** → **Watch Window** from the menu bar. This will bring up a **Watch Window** with local variables listed in the **Watch Locals** tab. To add a new expression to the **Watch Window**, select the **Watch 1** tab, then type point (or any expression you desire to examine) in the Name column. Then, choose **Debug** → **Run** or press F5. The program stops at the breakpoint and the **Watch Window** displays the value of the pointer. This is the value before the pointer is set to 0x80000000. By pressing F10 to step over the line, or the shortcut button ⏻, you should be able to see the value 0x80000000 in the **Watch Window**.

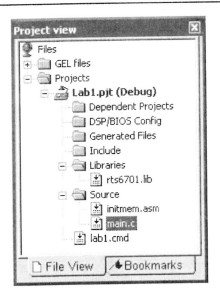

Figure 4-26: Project View panel.

To add a simple C function that sums the values, we can simply pass the pointer to the array and have a return type of integer. For the time being, what is of concern is not how the variables are passed, but rather how much time it takes to perform the operation.

The following simple function can be used to sum the values and return the result:

```c
#include <stdio.h>

void main()
{
    int i,ret;
    short *point;

    point = (short *) 0x80000000;

    printf("BEGIN\n");

    for(i=0;i<10;i++)
    {    printf("[%d] %d\n",i, point[i]);    }

    ret = ret_sum(point,10);

    printf("Sum = %d\n",ret);
    printf("END\n");
}

int ret_sum(const short* array,int N)
```

```
{
    int count,sum;
    sum = 0;

    for(count=0 ; count < N ; count++)
        sum += array[count];

    return(sum);
}
```

As part of the debugging process, it is normally required to benchmark or time the application program. In this lab, let us determine how much time it takes for the function ret_sum() to run. To achieve this benchmarking, reload the program and choose **Profiler → Start New Session**. This will bring up **Profile Session Name**. Type a session name, MySession by default, then click **OK**. The **Profile** window showing code size and statistics about the number of cycles will be docked at the bottom of CCS. Resize this window by dragging its edges or undock it so that you can see all the columns. Now right-click on the code inside the function to be benchmarked, then choose **Profile Function → in ... Session**. The name of the function will be added to the list in the **Profile** window. Finally, press F5 to run the program. Examine the number of cycles shown in Figure 4-27 for ret_sum(). It should be about 732 cycles (the exact number may slightly vary). This is the number of cycles it takes to execute the function ret_sum().

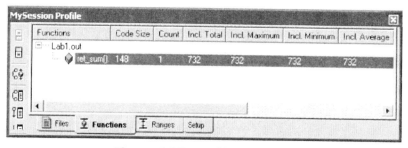

Figure 4-27: Profile window.

There is another way to benchmark codes using breakpoints. Double-click on the file *main.c* in the **Project View** panel and choose **View → Mixed Source/ASM** to list the assembled instructions corresponding to C code lines. Set a breakpoint at the calling line by placing the cursor on the line that reads ret = ret_sum(point,10), then press F9 or double-click **Selection Margin** located on the left-hand side of the editor. Set another breakpoint at the next line as indicated in Figure 4-28. Once the breakpoints are set, choose **Profiler → Enable Clock** to enable the profile clock. Then, choose **Profiler** → **View Clock** to bring up a window displaying **Profile Clock**. Now, press F5 to run the program. When the program is stopped at the first breakpoint, reset the clock by double-clicking the inner area of the **Profile Clock** window. Finally, click **Step Out** or **Run** in the **Debug** menu to execute and stop at the second breakpoint. Examine the number of clocks in the **Profile Clock** window. It should read 752. The discrepancy between the breakpoint and the profile approaches is originated from the extra procedures for calling functions, for example, passing arguments to function, storing return address, branching back from function, and so forth.

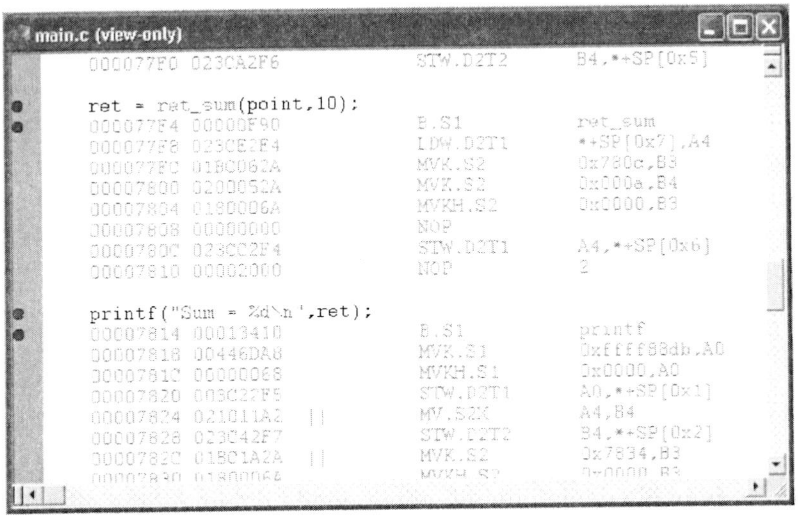

Figure 4-28: Profiling code execution time with breakpoint.

A workspace containing breakpoints, probe points, graphs, and watch windows, can be stored for recalling purposes. To do so, choose **File** → **Workspace** → **Save Workspace As**. This will bring up the **Save Work** window. Type the workspace name in the **File name** field, then click **Save**.

The file shown next is an assembly program for calculating the sum of the values. Here, the two arguments of the sum function are passed in registers A4 and B4. The return value gets stored in A4 and the return address in B3. The order in which the registers are chosen is governed by the passing argument convention discussed later. The name of the function should be preceded by an underscore as .global _sum. Create a new source file *sum.asm*, as shown next, and add it to the project so that main() can call the function sum().

```
        .global      _sum

_sum:
        ZERO    .L1     A9              ;Sum register
        MV      .L1     B4,A2           ;initialize counter with passed argument

loop:   LDH     .D1     *A4++, A7       ;load value pointed by A4 into reg. A7
        NOP             4
        ADD     .L1     A7,A9,A9        ;A9 += A7
 [A2]   SUB     .L1     A2,1,A2         ;decrement counter
 [A2]   B       .S1     loop            ;branch back to loop
        NOP             5

        MV      .L1     A9,A4           ;move result into return register A4
        B       .S2     B3              ;branch back to address stored in B3
        NOP             5
```

To save the file, go to the **Save as type** field and select **Assembly Source Files (*.asm)** from the pull-down list.

The program main() must also be modified by adding a function call to the assembly function sum(). This program is shown in Figure 4-29. Build the program and run it. You should be able to see the same return value.

```
#include <stdio.h>

void main()
{
     int i,ret;
     short *point;

     point = (short *) 0x80000000;

     printf("BEGIN\n");

     for(i=0;i<10;i++)
     {      printf("[%d] %d\n",i, point[i]);      }

     ret = ret_sum(point,10);
     printf("C program Sum = %d\n",ret);

     ret = sum(point,10);
     printf("Assembly program Sum = %d\n", ret);
     printf("END\n");
}

int ret_sum(const short* array,int N)
```

```
int ret_sum(const short* array,int N)
{
    int count,sum;
    sum = 0;

    for(count=0 ; count < N ; count++)
        sum += array[count];

    return(sum);
}
```

Figure 4-29: Lab1 complete program.

Table 4-2 provides the number of cycles it takes to run the sum function using several different builds. When a program is too big to fit into the internal memory, it has to be placed into the external memory. Although the program in this lab is small enough to fit in the internal memory, it is placed in the external memory to study the change in the number of cycles. To move the program into the external memory, open the *lab1.cmd* file and replace the line .text > IRAM with .text > CE0. As seen in Table 4-2, this build slows down the execution to 2535 cycles. In the second build, the program resides in the internal memory and the number of cycles is hence reduced to 732. By increasing the optimization level, the number of cycles can be further decreased to 281. The assembly version of the program gives 472 cycles. This is slower than the fully optimized C program because it is not yet optimized. Optimization of assembly codes will be discussed later. At this point, it is worth pointing out that, throughout the labs, the stated numbers of cycles indicate timings on the C6711 DSK with CCS version 2.2. The numbers of cycles will vary slightly depending on the DSK target and CCS version used.

Table 4-2: Number of cycles for different builds (on C6711 DSK with CCS2.2).

| | | Number of Cycles | |
Type of Build	Code size	Data in external memory	Data in internal memory
C program in external memory	148	2535	2075
C program in internal memory	148	732	382
–o0	64	464	113
–o1	64	463	111
–o2	100	404	57
–o3	84	281	33
Assembly program	64	472	150

L1.3 EVM Target

As described in Section 4.1 and illustrated in Figure 4-7, the memory map of EVM possesses more internal/external memory space than that of DSK, thus allowing more flexibility in loading code into the program/data section. Figure 4-30 shows an example using the EVM target where data located at EXT3, say 0x03000000, is placed at the beginning of EXT3.

```
MEMORY
{
    PMEM          : origin = 0x00000000,   length = 0x00010000
    EXT2          : origin = 0x02000000,   length = 0x01000000
    EXT3          : origin = 0x03000000,   length = 0x01000000
    DMEM          : origin = 0x80000000,   length = 0x00010000
}


SECTIONS
{
        .vectors    >  PMEM
        .text       >  PMEM
        .bss        >  DMEM
        .cinit      >  DMEM
        .const      >  DMEM
        .stack      >  DMEM
        .cio        >  DMEM
        .sysmem     >  DMEM
        .far        >  EXT2
        .mydata     >  EXT3
}
```

Figure 4-30: Command file for Lab 1.

In order to place the program code into the external memory, replace the line `.text > PMEM` with `.text > EXT2`. The number of cycles on the EVM target is shown in Table 4-3. To build the project, the library *rts6701.lib* for C6701 EVM or *rts6201.lib* for C6201 EVM needs to be added into the project. The **Target Version** field of the **Build option** should be selected as **C670x (–mv 6700)** for the floating-point target DSP TMS320C6701 EVM or **C620x (–mv 6200)** for the fixed-point target DSP TMS320C6201 EVM.

As a final note, it is worthwhile to mention a remark about the EVM board resetting. Sometimes you may notice that your program cannot be loaded into the DSP even though there is nothing wrong with it. Under such circumstances, you need to reset the EVM board to fix the problem. However, you have to close CCS before you reset the board. Otherwise, the problem will not be resolved.

Table 4-3: Number of cycles for different builds (on C6701 EVM with CCS2.2).

Type of Build	Code size	Number of Cycles	
		Data in external memory	Data in internal memory
C program in external memory	148	1185	955
C program in internal memory	148	541	355
-o0	64	282	97
-o1	64	281	95
-o2	100	213	36
-o3	84	129	21
Assembly program	64	139	133

L1.4 Simulator

When no DSP board is available, the CCS simulator can be used to run the lab programs. To configure CCS as a simulator, simply select **simulator** in the field **Platform** during the installation process of CCS, as shown in Figure 4-31. In the **Import Configurations** window, select the simulator option for one of the specified DSP boards. By clicking the button **Import**, then the button **Save and Quit**, the simulator gets configured and becomes ready to use. Note that although the simulator supports DMA and EMIF operations, operations related to McBSP, HPI and Timer are not supported. The files for running the labs via the simulator are provided under the simulator folder on the accompanying CD-ROM.

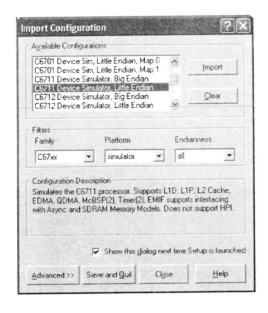

Figure 4-31: Simulator installation.

Bibliography

[1] Texas Instruments, *TMS320C6000 Code Composer Studio Tutorial*, Literature ID# SPRU 301C, 2000.

[2] Texas Instruments, *Code Composer Studio User's Guide*, Literature ID# SPRU 328B, 2000.

CHAPTER 5

Interrupt Data Processing

On a DSP processor, the processing of samples can be done within an ISR (interrupt service routine). Let us first discuss interrupts. As the name implies, an interrupt causes the processor to halt whatever it is processing in order to execute an ISR. An interrupt can be issued externally or internally. Twelve CPU interrupts are available on the C6x processor. The priorities of these interrupts are shown in Table 5-1. RESET is the highest priority interrupt. It halts the CPU and initializes all the registers to their default values. Non-maskable interrupt (NMI) is used for non-maskable or uninterruptible processing provided that the NMIE bit of the control register CSR (control status register) is set to zero. As indicated in Figure 5-1, there are a total of 16 interrupt sources while there are only 12 CPU interrupts. As a result, an interrupt source must be mapped to a CPU interrupt. This is done by setting appropriate bits of the two memory mapped Interrupt Multiplex registers.

Interrupt Name	Priority
RESET	Highest
NMI	
INT4	
INT5	
INT6	
INT7	
INT8	
INT9	
INT10	
INT11	
INT12	
INT13	
INT14	
INT15	Lowest

**Table 5-1:
Priorities of CPU interrupts.**

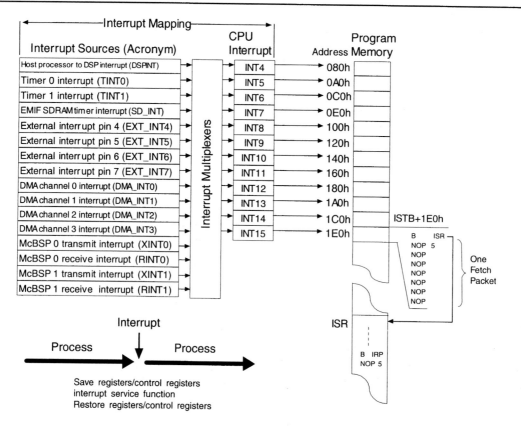

Figure 5-1: Interrupt mapping and operation.

Interrupts can be enabled or disabled by setting or clearing appropriate bits in interrupt enable register (IER). There is a master switch, the global interrupt enable (GIE) bit as part of CSR, which can be used to turn all interrupts on or off. For example, the assembly code shown in Figure 5-2 indicates how to enable INT4 and the GIE bit. Here the instruction MVC (move to and from a control register) is used to transfer a control register to a CPU register for bit manipulation. Another register called interrupt flag register (IFR) allows one to check if or what interrupt has occurred. (Refer to the *TI TMS320C6x CPU* manual [1] for more details on the interrupt registers.)

```
MVK     .S2     0010h, B3           ; bit4="1"
MVC     .S2     IER, B4             ; get IER
OR      .L2     B3, B4, B4          ; set bit4
MVC     .S2     B4, IER             ; write IER
MVC     .S2     CSR, B5             ; get CSR_GIE
OR      .L2     1, B5, B5           ; bit0="1"
MVC     .S2     B5, CSR             ; set GIE
```

Figure 5-2: Setup code to turn on INT4 and GIE.

The location where the processor will go to after an interrupt occurs is specified by a predefined offset for that interrupt added to the interrupt service table base (ISTB) bits as part of the interrupt service table pointer (ISTP) register. As an example, for the CPU INT15, the processor goes to the location ISTB + 1E0h. At this location, there is normally a branch instruction that would take the processor to a receive ISR somewhere in memory, as shown in Figure 5-1.

In general, an ISR includes three parts. The first and last part incorporate saving and restoring registers, respectively. The actual interrupt routine makes up the second part. If needed, saving and restoring are done to bring the status of the processor back to the time when the interrupt was issued.

Bibliography

[1] Texas Instruments, *TMS320C6000 CPU and Instruction Set Reference Guide*, Literature ID# SPRU 189F, 2000.

Lab 2: Audio Signal Sampling

The purpose of this lab is to use a C6x DSK or EVM to sample an analog audio signal in real-time. A common approach to processing live signals, which is the use of an interrupt service routine, is utilized here.

The AD535 codec on the C6711 DSK board provides a fixed sampling rate of 8 kHz. As done here, the audio daughter card PCM3003, shown in Figure 5-3, can be added to the C6711 DSK board in order to alter the sampling rate. PCM3003 has two stereo 3.5 mm audio jacks for a line-in and a line-out signal, plus two on-board microphones. The sampling rate can be varied from 4 kHz to 48 kHz via its timer. On the C6701 EVM board, there exists a 16-bit stereo audio codec CS4231A which can handle sampling rates from 5.5 kHz to 48 kHz. There are three 3.5 mm audio jack inputs on the back of the EVM board for a microphone-in, a line-in and a line-out signal. Each audio jack has its own amplifying and filtering capabilities. The block diagrams of the DSK and EVM stereo interface are shown in Figure 5-3. The codecs on both boards are connected to the C6x DSP through the multichannel buffered serial port (McBSP).

The configuration and control of the peripherals are done via application programming interface (API) functions of the chip support library (CSL) library for C6711 DSK, the board support library (BSL) for C6713 and C6416 DSK, and DSP support software for EVM. In this lab, CSL without DSP/BIOS is used to perform audio signal sampling. In Chapters 9 and 10, CSL is used with DSP/BIOS for real-time analysis and scheduling.

Considering that the peripherals and thus the libraries for the DSK and the EVM boards are different, the hardware and software configurations for these boards are presented in separate sections.

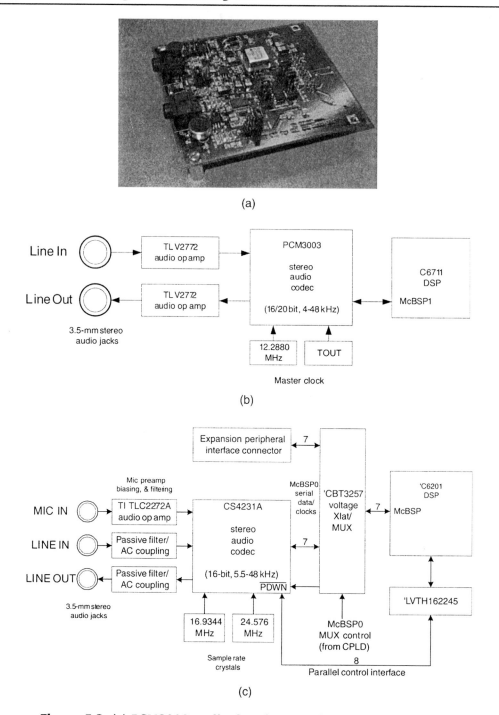

(a)

(b)

(c)

Figure 5-3: (a) PCM3003 audio daughter card†,(b) DSK stereo audio daughter card interface, (c) EVM stereo audio interface†.

L2.1 Initialization of Codec and Peripherals of C6711

In writing a program that uses the codec to sample an incoming analog signal, several initializations have to be performed. Among these are the initialization of the CSL library, McBSP, and timer. To achieve these initializations or adjustments, the API functions are used. Once the required initializations are made, an interrupt needs to be assigned to the receive register of the serial port to halt the processor and jump to a defined interrupt service routine. The final program will output the same input sample back to the codec. The following program includes an order of API functions that achieves all of the foregoing mentioned initializations (Figure 5-4 shows the flowchart of the steps involved):

```c
#define CHIP_6711          // Chip Symbol

#include <stdio.h>
#include <c6x.h>
#include <csl.h>
#include <csl_mcbsp.h>
#include <csl_irq.h>

MCBSP_Handle hMcbsp;

void hookint(void);
void interrupt serialPortRcvISR(void);

int main()
{
    MCBSP_Config MyConfig = {
        0x00010001,    // (SPCR) Enable serial port tx & rx
        0x000000A0,    // (RCR)  1-phase 32 bits receive
        0x000000A0,    // (XCR)  1-phase 32 bits transmit
        0x00000000,    // (SRGR)
        0x00000000,    // (MCR)
        0x00000000,    // (RCER)
        0x00000000,    // (XCER)
        0x00000000     // (PCR)
    };

    TIMER_Config timerCfg = {
        0x000003C1,    // (CTL) Internal clock src & TOUT is timer output
        0x00000000,    // (PRD) Fs = 73,242 Hz
        0x00000000     // (CNT)
    };

    TIMER_Handle hTimer;

    CSL_init();                    // Initialize the library

    hMcbsp = MCBSP_open(MCBSP_DEV1,MCBSP_OPEN_RESET);
    if (hMcbsp == INV)  {
        printf("Error opening MCBSP 1\n");
        return(-1);
    }
```

```
        MCBSP_config(hMcbsp,&MyConfig);

        hTimer = TIMER_open(TIMER_DEV0, TIMER_OPEN_RESET);
        if (hTimer == INV)   {
                printf("Error opening TIMER\n");
                return(-1);
        }
        TIMER_config(hTimer, &timerCfg);

        hookint();

        while(1)
        {
        }
}

void hookint()
{
        IRQ_globalDisable();            // Globally disables interrupts
        IRQ_nmiEnable();                // Enables the NMI interrupt
        IRQ_map(IRQ_EVT_RINT1,15);      // Maps an event to intr 15
        IRQ_enable(IRQ_EVT_RINT1);      // Enables the event
        IRQ_globalEnable();             // Globally enables intrs
}

interrupt void serialPortRcvISR()
{
        int temp;

        temp = MCBSP_read(hMcbsp);
        MCBSP_write(hMcbsp,temp);
}
```

Let us explain this program in a step-by-step fashion. Here, CSL is used without DSP/BIOS. This requires defining the chip identification symbol either in the source code or build option. The first line in the code specifies the chip identification symbol CHIP_6711 for the C6711 DSK. One can also define this symbol by stating CHIP_6711 in the field **Compiler tab** → **Preprocessor category** → **Define Symbols** in the menu **Project** → **Build Options**.

The first step consists of initializing the CSL library. This is done by using the function csl_init(), which must be called at the beginning of the program before calling any other CSL API functions.

Next step involves opening a handle to the McBSP in order to send and receive data. The McBSP API functions are used for this purpose. The API function MCBSP_ open() opens the McBSP, and returns the device handle hMcbsp for controlling the McBSP. The first argument of this function represents the port to be opened. Since the audio daughter card is connected to port 1, the port number is specified as

MCBSP_DEV1. The second argument, MCBSP_OPEN_RESET, specifies the initialization of the port register based on the power-on defaults. It also disables and clears any associated interrupts. If this opening fails, the symbolic constant INV is returned (for more details, refer to the *TMS320C6000 Chip Support Library API Reference Guide* [1]).

Next, it is required to adjust the parameters of the McBSP. In the serial port control register (SPCR), the RRST and XRST fields are set to 1, so that the serial port receiving and transmitting capabilities are enabled. The RINTM field is configured in order to generate receive interrupts (RINT1), i.e. data gets in the data receive register (DRR) as a result of the codec sampling. The frame and word lengths are set to 0 and 32 bit in the receive control register (RCR) and transmit control register (XCR), respectively. The sample rate generator register (SRGR) controls frame period, frame length, and sample rate clock divider. The multichannel control register (MCR), receive channel enable register (RCER), and transmit channel enable register (XCER) are used to configure subframe data receive and transmit modes. And the pin control register (PCR) is used for general-purpose I/O configuration.

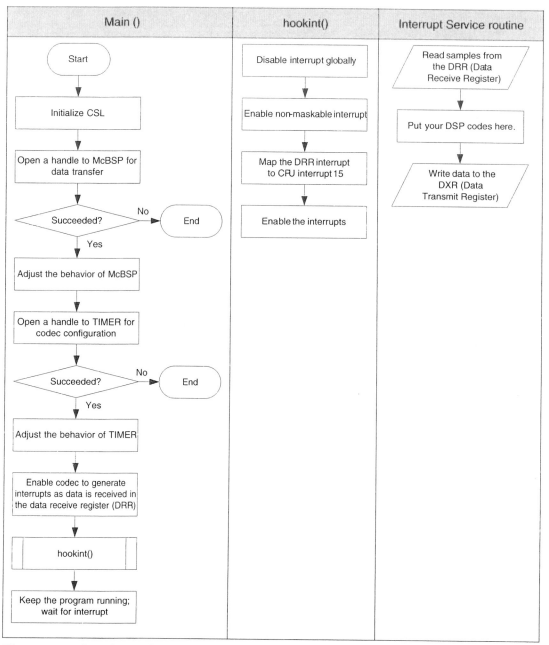

Figure 5-4: Flowchart of sampling program on C6711 DSK with audio daughter card.

The following lines of code perform the above configurations:

```
/*****************************************************/
/* configure McBSP                                   */
/*****************************************************/

MCBSP_Config MyConfig = {
    0x00010001,    // (SPCR)    Enable serial port tx & rx
    0x000000A0,    // (RCR)     1-phase 32 bits receive
    0x000000A0,    // (XCR)     1-phase 32 bits transmit
    0x00000000,    // (SRGR)
    0x00000000,    // (MCR)
    0x00000000,    // (RCER)
    0x00000000,    // (XCER)
    0x00000000     // (PCR)
};

hMcbsp = MCBSP_open(MCBSP_DEV1,MCBSP_OPEN_RESET);

if (hMcbsp == INV)
{
    printf("Error opening MCBSP 1\n");
    return(-1);
}

MCBSP_config(hMcbsp,&MyConfig);
```

Next, we need to adjust the codec parameters. Considering that the sampling rate of the audio daughter card is set to 48 kHz by default, its jumper setting needs to be changed in order to change the sampling rate by software. There are twelve jumpers on the daughter card for configuring data format, bit rate, data rate, enable/disable on-board microphone inputs, and master clock source. JP5 jumper connects pins 3 and 4 by default generating the master clock (MCLK) from the 12.288 MHz ADC clock. This jumper needs to be moved to connect pins 1 and 2 so that MCLK can be provided by the DSK timer.

The timer needs to be configured to set the sampling rate. The steps involved to open and configure the timer is similar to the McBSP. First, the timer is opened with the TIMER_open() API to get a handle, similar to the procedure to get a handle to McBSP. The first argument is used to select the timer device. Here, TIMER_DEV0 is used to specify timer 0. The second argument, TIMER_OPEN_RESET, initilaizes the timer device register with the power-on defaults, and any associated interrupts is disabled or cleared. If this opening fails, INV is returned.

The timer parameters need to be specified. In the timer control register (CTL), the CLKSRC field selects clock source, and the CP field selects pulse/clock mode. The HLD and GO fields are set to 1 in order to enable and start counting. If the FUNC field is set to 1, the timer output pin TOUT is configured as timer output, otherwise it serves as a general-purpose output. For this lab, the FUNC field is set to 1. The value in the timer period register (PRD) indicates the number of clock cycles to count before sending out the output clock signal. The PRD values and corresponding sampling rates are listed in Table 5-2. The listed values correspond to default positions of the jumpers except for JP5. The timer counter register (CNT) is incremented when it is enabled to count. The following lines of code are used to initialize the codec as just described.

Table 5-2: Examples of sampling rates using PCM3003 stereo audio codec.

PRD value	fs (Hz)
0	73242.19
1	36621.09
2	18310.55
3	12207.03
4	9155.27
5	7324.22
6	6103.52
7	5231.58
8	4577.64
9	4069.01

```
TIMER_Config timerCfg = {
     0x000003C1,   // (CTL) Internal clock src & TOUT is timer output
     0x00000000,   // (PRD) Fs = 73242 Hz
     0x00000000    // (CNT)
};

TIMER_Handle hTimer;

hTimer = TIMER_open(TIMER_DEV0, TIMER_OPEN_RESET);

if (hTimer == INV)
{
     printf("Error opening TIMER\n");
     return(-1);
}

TIMER_config(hTimer, &timerCfg);
```

The initializations of the CSL, McBSP and timer are now complete. Next, let us turn our attention to setting up an interrupt to branch to a simple ISR in order to process an incoming signal.

L2.2 Interrupt Service Routine

The idea of using interrupts is commonly used for real-time data processing. This approach is widely used, since it eliminates the need for complicated synchronization schemes. In our case, the interrupt occurs when a new data sample arrives in the DRR of the serial port. The generated interrupt will branch to an ISR, which is then used to process the sample and send it back out. To do this, the interrupt capabilities of the DSK must be enabled and adjusted so that an unused interrupt is assigned to the DRR event of the serial port.

The first task at hand is to initialize the interrupt service table pointer (ISTP) register with the base address of the interrupt service table (IST). Upon resetting the board, address 0 is assigned as the base address of the vector table. In this lab, we use this default value for ISTP, noting that the address can be relocated by changing the ISTB field value in the ISTP register.

Next, we need to select an interrupt source and map it to a CPU interrupt, in our case the McBSP1 receive interrupt (RINT1). Here, the CPU interrupt 15 is used and mapped to the RINT1 interrupt by using the function IRQ_map(). To connect the ISR to this interrupt, the IST needs to be modified. Let us define the ISR to be serialPortRcvISR(). The following assembly code defines the IST which hooks the CPU interrupt 15 to the ISR serialPortRcvISR().

```
; vectors.asm
    .ref    _c_int00
    .ref    _serialPortRcvISR ; refer the addr of ISR defined in C code
    .sect   "vectors"

    RESET_RST:  MVKL .S2 _c_int00, B0
                MVKH .S2 _c_int00, B0
                B    .S2 B0
                NOP
                NOP
                NOP
                NOP
                NOP
    NMI_RST:    .loop 8
                NOP
                .endloop
    RESV1:      .loop 8
                NOP
                .endloop
    RESV2:      .loop 8
                NOP
                .endloop
    INT4:       .loop 8
```

```
                       NOP
                       .endloop
         INT5:         .loop 8
                       NOP
                       .endloop
         INT6:         .loop 8
                       NOP
                       .endloop
         INT7:         .loop 8
                       NOP
                       .endloop
         INT8:         .loop 8
                       NOP
                       .endloop
         INT9:         .loop 8
                       NOP
                       .endloop
         INT10:        .loop 8
                       NOP
                       .endloop
         INT11:        .loop 8
                       NOP
                       .endloop
         INT12:        .loop 8
                       NOP
                       .endloop
         INT13:        .loop 8
                       NOP
                       .endloop
         INT14:        .loop 8
                       NOP
                       .endloop
         INT15:        MVKL .S2 _serialPortRcvISR, B0
                       MVKH .S2 _serialPortRcvISR, B0
                       B    .S2 B0              ;branch to ISR
                       NOP
                       NOP
                       NOP
                       NOP
                       NOP
```

Figure 5-5: Assembly code defining interrupt service table.

The last item to take care of is to enable interrupts by using the IRQ_enable and IRQ_globalEnable APIs. The following lines of code maps the CPU interrupt 15 to the RINT1 interrupt.

```
void hookint()
{
    IRQ_globalDisable();            // Globally disables interrupts
    IRQ_nmiEnable();                // Enables the NMI interrupt
    IRQ_map(IRQ_EVT_RINT1,15);      // Maps an event to intr 15
    IRQ_enable(IRQ_EVT_RINT1);      // Enables the event
    IRQ_globalEnable();             // Globally enables interrupts
}
```

A simple ISR can now be written to receive samples from the McBSP and send them back out, unprocessed for the time being. To write such an ISR, we need to state an interrupt declaration with no arguments. The MCBSP_read and MCBSP_write APIs are used to read samples from the DRR and write them to the DXR (data transmit register) of the McBSP. The device handler acquired during the configuration of the McBSP should be specified as an argument in both MCBSP_read and MCBSP_write. The ISR is presented below.

```
interrupt void serialPortRcvISR()
{
    int temp;

    temp = MCBSP_read(hMcbsp);
    MCBSP_write(hMcbsp,temp);
}
```

Considering that the CPU is not actually doing anything as it waits for a new data sample, an infinite loop is set up inside the main program to keep it running. As an interrupt occurs, the program branches to the ISR, performs it and then returns to its wait state. This is accomplished via a while(1){} statement.

Now the complete program for sampling an analog signal is ready for use. Basically, this program services interrupts to read in samples of an analog signal, such as the output from a CD player connected to the line-in of the DSK.

To build this program in CCS, the project should include two libraries: *rts6700.lib* and *csl6711.lib*. The library *rts6700.lib* is the runtime-support library containing the run-time support functions such as math functions. The chip support library *csl6711.lib* is a collection of the API modules for programming the registers and peripherals. This library allows the programmer to control interrupt functionality, CPU operational modes, and internal peripherals including McBSPs and timers. In addition to these library files, a linker command file needs to be added into the project. The following command file is used in this lab:

```
MEMORY
{
    vecs:        o = 00000000h    l = 00000200h
    IRAM:        o = 00000200h    l = 0000FE00h
    CE0:         o = 80000000h    l = 01000000h
}
```

```
SECTIONS
{
    "vectors"    >       vecs
    .cinit       >       IRAM
    .text        >       IRAM
    .stack       >       IRAM
    .bss         >       IRAM
    .const       >       IRAM
    .data        >       IRAM
    .far         >       IRAM
    .switch      >       IRAM
    .sysmem      >       IRAM
    .tables      >       IRAM
    .cio         >       IRAM
}
```

Figure 5-6 shows the **Project View** panel after the necessary files are added into the project. To build an executable file from these files, the button **Rebuild All** needs to be clicked. The executable file can then get loaded by choosing the menu item **File → Load Program**. By running the executable and connecting the output of a CD player to the line-in and a pair of powered speakers to the line-out, CD quality sound should be heard. Figure 5-7 shows the block diagram of this setup.

Figure 5-6: Project view for Lab 2.

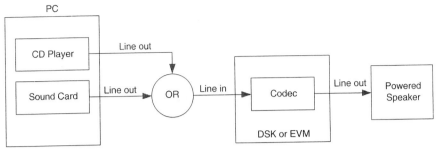

Figure 5-7: Block diagram of Lab 2 setup.

The effect of the sampling rate on the sound quality can be studied by modifying the value in the PRD register as part of the main program `codec.c` as indicated below:

```
int main()
{
    . . .
    TIMER_Config timerCfg = {
        0x000003C1,   // (CTL) Internal clock src & TOUT is a timer output
        0x00000009,   // (PRD) Fs = 4069 Hz
        0x00000000    // (CNT)
    };
}
```

This will change the sampling rate to 4 kHz. By rebuilding, reloading, and running the executable code, degradation in the sound quality can be heard due to the deviation from the Nyquist rate.

It is possible to manipulate or process the audio signal. For example, the sound volume can be controlled by multiplying a volume gain factor with the sound samples. The code for doing so is as follows:

```
int volumeGain;

int main()
{
    . . .
    volumeGain = 1; /* Initialize */
    . . .
}
interrupt void serialPortRcvISR()
{
    int temp;

    temp = MCBSP_read(hMcbsp);
    temp = temp * volumeGain;
    MCBSP_write(hMcbsp, temp);
}
```

The variable `volumeGain` is declared as a global variable in order to be accessed at run-time. To change the volume at run-time, the option **Edit** → **Variable** should be chosen, which brings up an **Edit Variable** dialog box. As shown in Figure 5-8, by entering `volumeGain` in the **Variable** field and a desired gain value in the **Value** field of this dialog box, the sound volume can be altered.

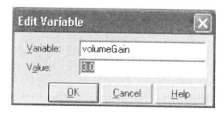

Figure 5-8: Editing value of a variable.

L2.3 C6711 DSK without Audio Daughter Card

If the codec AD535 on the DSK board is used instead of the daughter card, the McBSP, codec, interrupt configurations need to be modified, considering that the codec sampling rate is fixed at 8 kHz. The timer configuration is no longer needed. The McBSP configuration stays the same as before with one difference in the register value corresponding to the codec data format. The codec configuration is done based on the secondary communication mode of AD535 via McBSP0. That is, the LSB of the data must be first set to 1 to get the codec ready for configuration. Then, during the primary communication mode, the LSB needs to remain 0 for general purpose I/O. This is done by performing a masking operation with `0xFFFE`. The AD535 codec is configured to select TAPI & microphone preamps for ADC input, set voice ADC input & DAC output PGA gain as 0 dB, and set 60-ohm speaker L/R buffer gain as 0 dB. The interrupt configuration is the same as before except for the interrupt source being `IRQ_EVT_RINT0`, since the codec AD535 is accessed via the McBSP channel 0.

The source code of the foregoing configurations is shown below.

```
#define CHIP_6711

#include <stdio.h>
#include <c6x.h>
#include <csl.h>
#include <csl_mcbsp.h>
#include <csl_irq.h>
```

```
/*****************************************************************
 *  Declarations
 *****************************************************************/

MCBSP_Handle hMcbsp;
int    volumeGain;

interrupt void serialPortRcvISR(void);
void AD535_Init(int data);

void main()
{
    MCBSP_Config MyConfig = {
        0x00012001,         // (SPCR)  Serial port tx & rx are enabled.
        0x00010040,         // (RCR)   one 16 bit data/frame
        0x00010040,         // (XCR)   one 16 bit data/frame
        0x00000000,         // (SRGR)
        0x00000000,         // (MCR)
        0x00000000,         // (RCER)
        0x00000000,         // (XCER)
        0x00000000          // (PCR)
    };

    CSL_init();             // Initialize the library

    hMcbsp = MCBSP_open(MCBSP_DEV0, MCBSP_OPEN_RESET);
    MCBSP_config(hMcbsp,&MyConfig);

    AD535_Init(0x0306); // Reg 3, Voice channel software reset not asserted
                        // TAPI & Microphone preamps selected for ADC input
    AD535_Init(0x0400); // Reg 4, Voice ADC input PGA gain = 0 dB
    AD535_Init(0x0502); // Reg 5, Voice DAC output PGA gain = 0 dB
                        // 60-Ohm Spkr_L/R buffer gain = 0 dB

    volumeGain = 1;

    IRQ_globalDisable();            // Globally disables interrupts
    IRQ_nmiEnable();                // Enables the NMI interrupt
    IRQ_map(IRQ_EVT_RINT0,15);      // Maps an event to a physical interrupt
    IRQ_enable(IRQ_EVT_RINT0);      // Enables the event
    IRQ_globalEnable();             // Globally enables interrupts

    while(1)
    {
    }
}

void AD535_Init(int data)
{
    while(!MCBSP_xrdy(hMcbsp));  // Secondary serial communication request
    MCBSP_write(hMcbsp, 1);

    while(!MCBSP_rrdy(hMcbsp));  // Read
    MCBSP_read(hMcbsp);
```

```
        while(!MCBSP_xrdy(hMcbsp));     // Write to Control Register of AD535
        MCBSP_write(hMcbsp, data);

        while(!MCBSP_rrdy(hMcbsp));     // Read
        MCBSP_read(hMcbsp);
}

interrupt void serialPortRcvISR()
{
        int temp;

        temp = MCBSP_read(hMcbsp);
        temp = ( temp * volumeGain ) & 0xFFFE;

    MCBSP_write(hMcbsp,temp);
}
```

L2.4 C6416/C6713 DSK

The codec AIC23 on the C6416/C6713 DSK is different than the codec AD535 on the C6711 DSK. The configuration of this codec is achieved by using the C6416/C6713 Board Support Library (BSL) as part of CCS. Two serial channels, McBSP1 and McBSP2 (McBSP0 and McBSP1 for C6713), are used for the configuration. The McBSP1 (McBSP0 for C6713) is used to control the codec internal configuration registers. The McBSP2 (McBSP1 for C6713) is used for audio data communication. Consequently, the interrupt source should be stated as IRQ_EVT_RINT2 (IRQ_EVT_RINT1 for C6713).

The sampling rates supported by the AIC23 codec are listed in Table 5-3. The sampling rate is configured using the DSK6416_AIC23_setFreq() (DSK6713_AIC23_setFreq() for C6713) API. Using the default configuration of AIC23, the sampled data is stored in a frame consisting of 16-bit left channel data followed by 16-bit right channel data. In order to have code consistency across different DSP platforms, the default configuration of McBSP defined in the DSK6416_AIC23 library needs to be customized. This is achieved by modifying the transmit/receive interrupt mode (TINTM/RINTM) field of SPCR so that the transmit/receive interrupt (XINT/RINT) is generated by a new frame synchronization. Also, the transmit/receive word length should be set to 32 bits to process data from left and right together. A more detailed description of the codec internal registers is provided in [3].

Table 5-3: Sampling rates allowed by AIC23 stereo audio codec.

Sampling rate (kHz)	
8.000	44.000
16.000	48.000
24.000	96.000
32.000	

To build the project, the Board Support Library, *dsk6416bsl.lib* (*dsk6713bsl.lib* for C6713), needs to be added to the project. The source code of the foregoing configurations for C6416 is shown below.

```
#define CHIP_6416

#include <stdio.h>
#include <c6x.h>
#include <csl.h>
#include <csl_mcbsp.h>
#include <csl_irq.h>

#include "dsk6416.h"
#include "dsk6416_aic23.h"

DSK6416_AIC23_CodecHandle hCodec;
DSK6416_AIC23_Config config = DSK6416_AIC23_DEFAULTCONFIG;
                        // Codec configuration with default settings

interrupt void serialPortRcvISR(void);
void hook_int();
int volumeGain;

void main()
{
    DSK6416_init();          // Initialize the board support library
    hCodec = DSK6416_AIC23_openCodec(0, &config);

    MCBSP_FSETS(SPCR2, RINTM, FRM);
    MCBSP_FSETS(SPCR2, XINTM, FRM);
    MCBSP_FSETS(RCR2, RWDLEN1, 32BIT);
    MCBSP_FSETS(XCR2, XWDLEN1, 32BIT);

    DSK6416_AIC23_setFreq(hCodec, DSK6416_AIC23_FREQ_48KHZ);

    volumeGain = 1;

    hook_int();

    while(1)
    {
```

```
        }
}

void hook_int()
{
        IRQ_globalDisable();            // Globally disables interrupts
        IRQ_nmiEnable();                // Enables the NMI interrupt
        IRQ_map(IRQ_EVT_RINT2,15);      // Maps an event to a physical interrupt
        IRQ_enable(IRQ_EVT_RINT2);      // Enables the event
        IRQ_globalEnable();             // Globally enables interrupts
}

interrupt void serialPortRcvISR()
{
        Uint32 temp;

        temp = MCBSP_read(DSK6416_AIC23_DATAHANDLE);
        temp = temp * volumeGain;
        MCBSP_write(DSK6416_AIC23_DATAHANDLE, temp);
}
```

L2.5 C67x EVM

All the configurations stated for the DSK can be done in a similar way for the C67x
EVM through the DSP support software provided by Texas Instruments. The DSP
support software contains C functions for accessing and setting up the EVM board,
McBSP, and codec. The codec library is supplied in the archived object library file
drv6x.lib (*drv6xe.lib* is the big-endian version of this library). The corresponding
source file is *drv6x.src*. The codec library contains API functions that can be used
to configure and control the operation of the codec. The functional descriptions of
these functions can be found in the *EVM Reference Guide* [2] under TMS320C6x
EVM DSP Support Software. These functions are utilized here to write a sampling
program for the C67x EVM.

Since the main structure of the code for EVM is exactly the same as that for DSK,
this section includes only the parts that differ with DSK. The EVM version of the
program is shown below.

```
#include <stdlib.h>
#include <stdio.h>
#include <string.h>
#include <common.h>
#include <mcbspdrv.h>
#include <intr.h>
#include <board.h>
#include <codec.h>
#include <mcbsp.h>
```

```
#include <mathf.h>

void hookint(void);
interrupt void serialPortRcvISR(void);

int main()
{
      Mcbsp_dev dev;
      Mcbsp_config  mcbspConfig;
      int sampleRate, status;

      /*********************************************************/
      /* Initialize EVM                                        */
      /*********************************************************/
      status = evm_init();
      if(status == ERROR)
            return (ERROR);

      /*********************************************************/
      /* Open MCBSP for subsequent Examples                    */
      /*********************************************************/
      mcbsp_drv_init();
      dev= mcbsp_open(0);
      if (dev == NULL)
      {
            printf("Error opening MCBSP 0\n");
            return(ERROR);
      }

      /*********************************************************/
      /* configure McBSP                                       */
      /*********************************************************/
      memset(&mcbspConfig,0,sizeof(mcbspConfig));
      mcbspConfig.loopback              = FALSE;
      mcbspConfig.tx.update             = TRUE;
      mcbspConfig.tx.clock_mode         = CLK_MODE_EXT;
      mcbspConfig.tx.frame_length1      = 0;
      mcbspConfig.tx.word_length1       = WORD_LENGTH_32;
      mcbspConfig.rx.update             = TRUE;
      mcbspConfig.rx.clock_mode         = CLK_MODE_EXT;
      mcbspConfig.rx.frame_length1      = 0;
      mcbspConfig.rx.word_length1       = WORD_LENGTH_32;
      mcbsp_config(dev,&mcbspConfig);
      MCBSP_ENABLE(0, MCBSP_BOTH);

      /*********************************************************/
      /* configure CODEC                                       */
      /*********************************************************/
      codec_init();

      /* A/D 0.0 dB gain, turn off 20dB mic gain, sel (L/R)LINE input */
      codec_adc_control(LEFT,0.0,FALSE,LINE_SEL);
      codec_adc_control(RIGHT,0.0,FALSE,LINE_SEL);

      /* mute (L/R)LINE input to mixer          */
      codec_line_in_control(LEFT,MIN_AUX_LINE_GAIN,TRUE);
      codec_line_in_control(RIGHT,MIN_AUX_LINE_GAIN,TRUE);
```

```
        /* D/A 0.0 dB atten, do not mute DAC outputs */
        codec_dac_control(LEFT, 0.0, FALSE);
        codec_dac_control(RIGHT, 0.0, FALSE);


        sampleRate = 44100;
        codec_change_sample_rate(sampleRate, TRUE);
        codec_interrupt_enable();
        hookint();


        /******************************************************/
        /* Main Loop, wait for Interrupt                      */
        /******************************************************/

        while (1)
        {
        }
}

void hookint()
{
        intr_init();
        intr_map(CPU_INT15, ISN_RINT0);
        intr_hook(serialPortRcvISR, CPU_INT15);
        INTR_ENABLE(15);
        INTR_GLOBAL_ENABLE();
        return;
}

interrupt void serialPortRcvISR(void)
{
        int temp;
        temp = MCBSP_READ(0);
        MCBSP_WRITE(0, temp);
}
```

Similar to DSK, the first part of the program involves initialization. Since we are using the DSP support software instead of CSL, the EVM initialization is done by stating the function evm_init() before calling any other support functions. This function configures the EVM base address variables, and initializes the external memory interface (EMIF). The return value of this function indicates success or failure of the EVM initialization.

Once EVM has been successfully initialized, next step is to open a handle to the McBSP in order to send and receive data. The McBSP API functions are used for this purpose. The API function mcbsp_drv_init() initializes the McBSP driver and allocates memory for the device handles. The return value of this function also indicates success or failure. After the initialization of the McBSP driver, the data structure elements that control the behavior of the McBSP are set to their default values (for more details, refer to the *EVM Reference Guide* [2]). Then, the McBSP needs to be actually opened to get a handle to it. The API function mcbsp_open() is used to return the handle dev for controlling the McBSP.

Next step is to adjust the parameters of the McBSP. The data structure of the McBSP gets initialized to its default values as a result of using the initialization functions, so all that is required is the adjustment of several parameters to suit our needs. The loopback property of the McBSP is turned off or set to FALSE in order to disable the serial port test mode, in which the receive pins get connected internally to the transmit pins. The update property is set to TRUE for setting properties. The source signal for clocking the serial port transfers is made external by setting the clock mode to CLK_MODE_EXT. The frame and word lengths are set to 0 and WORD_LENGTH_32, respectively. The adjustments to the McBSP are made by allocating memory to the structure mcbsp_Config using the function memset(). The address of this structure is passed as an argument to the function mcbsp_config(), which performs the required adjustments.

Finally, the McBSP needs to be activated. This is done by using the macro MCBSP_ENABLE(), which is defined in the header file *mcbsp.h*. A macro is a collection of instructions that gets substituted for the macro in the program by the assembler. In this lab, the macro MCBSP_ENABLE(0) places the selected port 0 in the general purpose I/O mode. The following lines of code are used to do these adjustments:

```
memset(&mcbspConfig,0,sizeof(mcbspConfig));
mcbspConfig.loopback              = FALSE;
mcbspConfig.tx.update             = TRUE;
mcbspConfig.tx.clock_mode         = CLK_MODE_EXT;
mcbspConfig.tx.frame_length1      = 0;
mcbspConfig.tx.word_length1       = WORD_LENGTH_32;
mcbspConfig.rx.update             = TRUE;
mcbspConfig.rx.clock_mode         = CLK_MODE_EXT;
mcbspConfig.rx.frame_length1      = 0;
mcbspConfig.rx.word_length1       = WORD_LENGTH_32;
mcbsp_config(dev,&mcbspConfig);
MCBSP_ENABLE(0, MCBSP_BOTH);
```

Next, we need to adjust the parameters of the codec. The codec is initialized by using the codec API function codec_init(). This function sets the codec to its default parameters. The main item to adjust here is sampling rate. This is done by using the API function codec_change_sample_rate(). This function sets the sampling rate of the codec to the closest allowed sampling rate of the passed argument. The return value from this function will be the actual sampling rate. Table 5-4 lists the sampling rates supported by the codec. The other required adjustments are the selection of line-in or mic-in and the adjustment of their gain settings. To have stereo input, both channels should be selected and their gains adjusted to 0dB settings. The API functions that accomplish these tasks are `codec_adc_control()`, `codec_line_in_control()`, and `codec_dac_control()`. It is also required for the codec to generate interrupts as data is received in the DRR. Hence, the interrupt processing capability of the codec must be enabled. This is accomplished by using the API function `codec_interrupt_enable()`. The following lines of code are used for the purpose of initializing the codec as just described:

Table 5-4: Sampling rates allowed by CS4231A stereo audio codec.

Sampling rate (kHz)	
5.5125	22.0500
6.6150	27.4286
8.0000	32.0000
9.6000	33.0750
11.0250	37.8000
16.0000	44.1000
18.9000	48.0000

```
codec_init();

// ADC 0.0 dB gain, turn off 20dB mic gain, sel (L/R)LINE input
codec_adc_control(LEFT,0.0,FALSE,LINE_SEL);
codec_adc_control(RIGHT,0.0,FALSE,LINE_SEL);

// (L/R) LINE input to mixer
codec_line_in_control(LEFT,MIN_AUX_LINE_GAIN,FALSE);
codec_line_in_control(RIGHT,MIN_AUX_LINE_GAIN,FALSE);

// DAC 0.0 dB atten, do not mute DAC outputs
codec_dac_control(LEFT, 0.0, FALSE);
codec_dac_control(RIGHT, 0.0, FALSE);

sampleRate = 44100;
actualrate = codec_change_sample_rate(sampleRate, TRUE);
codec_interrupt_enable();
```

Now, in order to set up an interrupt, the first task involves the initialization of the interrupt service table pointer (ISTP) register with the address of the global `vec_table`, which is resolved at the link time. This is done by placing the base address of the vector table in the ISTP register. The function `intr_init()` is used for this purpose. Next, we need to select an interrupt number and map it to a CPU interrupt, in our case the RINT0 interrupt. Here, the CPU interrupt 15 is used and mapped to the RINT0 interrupt by using the function `intr_map()`. To connect an ISR to this interrupt, the function `intr_hook()` is called, to which the name of the function that we wish to use is passed. The last task is to enable the interrupts via the macros `INTR_ENABLE` and `INTR_GLOBAL_ENABLE`. The following lines of code map the CPU interrupt 15 to the RINT0 interrupt and then hook it to an ISR named `serialPortRcvISR`:

```
intr_init();
intr_map(CPU_INT15, ISN_RINT0);
intr_hook(serialPortRcvISR, CPU_INT15);
INTR_ENABLE(15);
INTR_GLOBAL_ENABLE();
```

To build this program in CCS, the project should include three libraries: *rts6701.lib*, *drv6x.lib* and *dev6x.lib*. The library *rts6701.lib* is the runtime-support library containing the runtime-support functions such as math functions. The library *dev6x.lib* is a collection of macros and functions for programming the C6x registers and peripherals. This library allows the programmer to control interrupt functionality, CPU operational modes, and internal peripherals including McBSPs. The linker command file for EVM is shown below.

```
MEMORY
{
  INT_PROG_MEM (RX)     : origin = 0x00000000 length = 0x00010000
  SBSRAM_PROG_MEM (RX)  : origin = 0x00400000 length = 0x00014000
  SBSRAM_DATA_MEM (RW)  : origin = 0x00414000 length = 0x0002C000
  SDRAM0_DATA_MEM (RW)  : origin = 0x02000000 length = 0x00400000
  SDRAM1_DATA_MEM (RW)  : origin = 0x03000000 length = 0x00400000
  INT_DATA_MEM (RW)     : origin = 0x80000000 length = 0x00010000
}

SECTIONS
{
  .vec:        load = 0x00000000
  .text:       load = SBSRAM_PROG_MEM
  .const:      load = INT_DATA_MEM
  .bss:        load = INT_DATA_MEM
  .data:       load = INT_DATA_MEM
  .cinit       load = INT_DATA_MEM
  .pinit       load = INT_DATA_MEM
  .stack       load = INT_DATA_MEM
  .far         load = INT_DATA_MEM
  .sysmem      load = SDRAM0_DATA_MEM
  .cio         load = INT_DATA_MEM
  sbsbuf       load = SBSRAM_DATA_MEM
               { _SbsramDataAddr = .; _SbsramDataSize = 0x0002C000; }
}
```

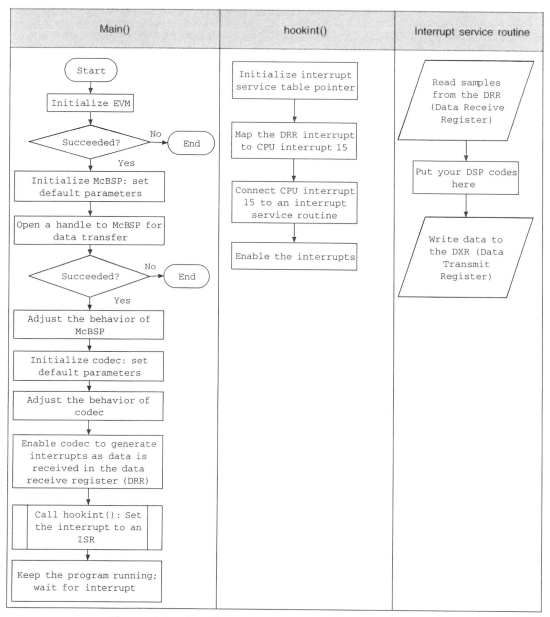

Figure 5-9: Flowchart of sampling program for EVM.

Bibliography

[1] Texas Instruments, *TMS320C6000 Chip Support Library API Reference Guide*, Literature ID# SPRU 401G, 2003.

[2] Texas Instruments, *TMS320C6201/6701 Evaluation Module User's Guide*, Literature ID# SPRU 269F, 2002.

[3] Texas Instruments, *Stereo Audio D/A Converter, 8-to 96-KHz, With Integrated Headphone Amplifier*, Literature ID# SLWS 106G, 2003.

[4] Spectrum Digital Inc., *TMS320C6416 DSK Technical Reference*, 2003.

CHAPTER 6

Fixed-Point vs. Floating-Point

One important feature that distinguishes different DSP processors is whether their CPUs perform fixed-point or floating-point arithmetic. In a fixed-point processor, numbers are represented and manipulated in integer format. In a floating-point processor, in addition to integer arithmetic, floating-point arithmetic can be handled. This means that numbers are represented by the combination of a mantissa (or a fractional part) and an exponent part, and the CPU possesses the necessary hardware for manipulating both of these parts. As a result, in general, floating-point processors are more expensive and slower than fixed-point ones.

In a fixed-point processor, one needs to be concerned with the dynamic range of numbers, since a much narrower range of numbers can be represented in integer format as compared to floating-point format. For most applications, such a concern can be virtually ignored when using a floating-point processor. Consequently, fixed-point processors usually demand more coding effort than do their floating-point counterparts.

6.1 Q-format Number Representation on Fixed-Point DSPs

The decimal value of a 2's-complement number $B = b_{N-1}b_{N-2}...b_1b_0$, $b_i \in \{0,1\}$, is given by

$$D(B) = -b_{N-1}2^{N-1} + b_{N-2}2^{N-2} + ... + b_1 2^1 + b_0 2^0 \tag{6.1}$$

2's-complement representation allows a processor to perform integer addition and subtraction by using the same hardware. When using unsigned integer representation, the sign bit is treated as an extra bit. This way only positive numbers can be represented.

There is a limitation to the dynamic range of the foregoing integer representation scheme. For example, in a 16-bit system, it is not possible to represent numbers larger than $+2^{15} - 1 = 32767$ and smaller than $-2^{15} = 32768$. To cope with this limitation, numbers are normalized between -1 and 1. In other words, they are represented as fractions. This normalization is achieved by the programmer moving the implied or imaginary binary point (note that there is no physical memory allocated to this point) as indicated in Figure 6-1. This way, the fractional value is given by

$$F(B) = -b_{N-1}2^0 + b_{N-2}2^{-1} + \ldots + b_1 2^{-(N-2)} + b_0 2^{-(N-1)} \qquad (6.2)$$

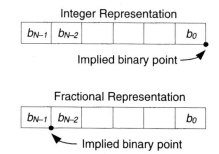

Integer Representation

Fractional Representation

Figure 6-1: Number representations.

This representation scheme is referred to as Q-format or fractional representation. The programmer needs to keep track of the implied binary point when manipulating Q-format numbers. For instance, let us consider two Q-15 format numbers, given that we have a 16-bit wide memory. Each number consists of 1 sign bit plus 15 fractional bits. When these numbers are multiplied, a Q-30 format number is obtained (the product of two fractions is still a fraction), with bit 31 being the sign bit and bit 32 another sign bit (called extended sign bit). If not enough bits are available to store all 32 bits, and only 16 bits can be stored, it makes sense to store the most significant bits. This translates into storing the upper portion of the 32-bit product register by doing a 15-bit right shift (SHR). In this manner, the product would be stored in Q-15 format. (See Figure 6-2.)

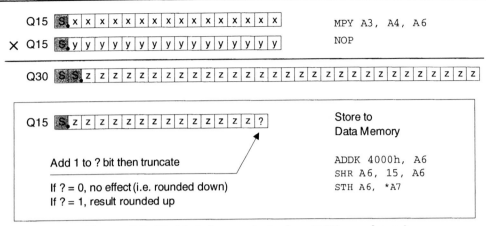

Figure 6-2: Multiplying and storing Q-15 numbers.†

Based on 2's-complement representation, a dynamic range of $-(2^{N-1}) \le D(B) < 2^{N-1} -1$ can be achieved, where N denotes the number of bits. For illustration purposes, let us consider a 4-bit system where the most negative number is –8 and the most positive number 7. The decimal representations of the numbers are shown in Figure 6-3. Notice how the numbers change from most positive to most negative with the sign bit. Since only the integer numbers falling within the limits –8 and 7 can be represented, it is easy to see that any multiplication or addition resulting in a number larger than 7 or smaller than –8 will cause overflow. For example, when 6 is multiplied by 2, we get 12. Hence, the result is greater than the representation limits and will be wrapped around the circle to 1100, which is –4.

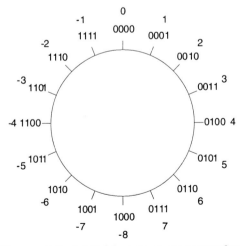

Figure 6-3: 4-bit binary representation.

Q-format representation solves this problem by normalizing the dynamic range between –1 and 1. Any resulting multiplication will be within the limits of this dynamic range. Using Q-format representation, the dynamic range is divided into 2^N sections, where $2^{-(N-1)}$ is the size of a section. The most negative number is always –1 and the most positive number is $1 - 2^{-(N-1)}$.

The following example helps one to see the difference in the two representation schemes. As shown in Figure 6-4, the multiplication of 0110 by 1110 in binary is the equivalent of multiplying 6 by –2 in decimal, giving an outcome of –12, a number exceeding the dynamic range of the 4-bit system. Based on the Q-3 representation, these numbers correspond to 0.75 and –0.25, respectively. The result is –0.1875, which falls within the fractional range. Notice that the hardware generates the same 1's and 0's, what is different is the interpretation of the bits.

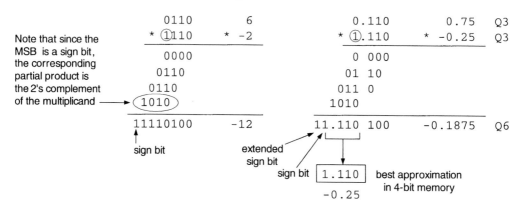

Figure 6-4: Binary and fractional multiplication.

When multiplying Q-N numbers, it should be remembered that the result will consist of 2N fractional bits, one sign bit, and one or more extended sign bits. Based on the datatype used, the result has to be shifted accordingly. If two Q-15 numbers are multiplied, the result will be 32-bits wide, with the MSB being the extended sign bit followed by the sign bit. The imaginary decimal point will be after the 30th bit. So a right shift of 15 is required to store the result in a 16-bit memory location as a Q-15 number. It should be realized that some precision is lost, of course, as a result of discarding the smaller fractional bits. Since only 16 bits can be stored, the shifting allows one to retain the higher precision fractional bits. If a 32-bit storage capability is available, a left shift of 1 can be done to remove the extended sign bit and store the result as a Q-31 number.

To further understand a possible precision loss when manipulating Q-format numbers, let us consider another example where two Q12 numbers corresponding to 7.5 and 7.25 are multiplied. As can be seen from Figure 6-5, the resulting product must be left shifted by 4 bits to store all the fractional bits corresponding to Q12 format. However, doing so results in a product value of 6.375, which is different than the correct value of 54.375. If the product is stored in a lower precision Q-format—say, in Q8 format—then the correct product value can be stored.

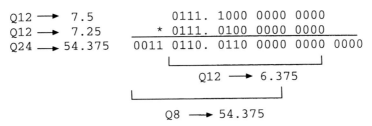

Figure 6-5: Q-format precision loss example.

Although Q-format solves the problem of overflow in multiplication, addition and subtraction still pose a problem. When adding two Q15 numbers, the sum exceeds the range of Q15 representation. To solve this problem, the scaling approach, discussed later in this chapter, needs to be employed.

6.2 Finite Word Length Effects on Fixed-Point DSPs

Due to the fact that memory or registers have finite number of bits, there could be a noticeable error between desired and actual outcomes on a fixed-point processor. The so-called finite word length quantization effect is similar to input data quantization effect introduced by an A/D converter.

Consider fractional numbers quantized by a $b + 1$ bit converter. When these numbers are manipulated and stored in an $M + 1$ bit memory, with $M < b$, there is going to be an error (simply because $b - M$ of the least significant fractional bits are discarded or truncated). This finite word length error could alter the behavior of a system to an unacceptable degree. The range of the magnitude of truncation error ε_t is given by $0 \le |\varepsilon_t| \le 2^M - 2^b$. The lowest level of truncation error corresponds to the situation when all the thrown-away bits are zeros, and the highest level to the situation when all the thrown-away bits are ones.

This effect has been extensively studied for FIR and IIR filters. (For example see [1].) Since the coefficients of such filters are represented by a finite number of bits, the roots of their transfer function polynomials, or the positions of their zeros and poles, shift in the complex plane. The amount of shift in the positions of poles and zeros can be related to the amount of quantization error in the coefficients. For example, for an Nth-order IIR filter, the sensitivity of the ith pole p_i with respect to the kth coefficient A_k can be derived to be (see [1]),

$$\frac{\partial p_i}{\partial A_k} = \frac{-p_i^{N-k}}{\displaystyle\prod_{\substack{l=1 \\ l \neq i}}^{N}(p_i - p_l)} \qquad (6.3)$$

This means that the change in the position of a pole is influenced by the positions of all the other poles. That is the reason the implementation of an Nth order IIR filter is normally achieved by having a number of second-order IIR filters in series in order to decouple this dependency of poles.

Also, note that as a result of coefficient quantization, the actual frequency response $\hat{H}(e^{j\theta})$ would become different than the desired frequency response $H(e^{j\theta})$. For example, for a FIR filter having N coefficients, it can be easily shown that the amount of error in the magnitude of the frequency response, $\left|\Delta H(e^{j\theta})\right|$, is bounded by

$$\left|\Delta H(e^{j\theta})\right| = \left|H(e^{j\theta}) - \hat{H}(e^{j\theta})\right| \leq N2^{-b} \qquad (6.4)$$

In addition to the above effects, coefficient quantization can lead to limit cycles. This means that in the absence of an input, the response of a supposedly stable system (poles inside the unit circle) to a unit sample is oscillatory instead of diminishing in magnitude.

6.3 Floating-Point Number Representation

Due to relatively limited dynamic ranges of fixed-point processors, when using such processors, one should be concerned with the scaling issue, or how big the numbers get in the manipulation of a signal. Scaling is not an issue when using floating-point processors, since the floating-point hardware provides a much wider dynamic range. The C67x processor is the floating-point version of the C6x family with many additional floating-point instructions. [Appendix A (Quick Reference Guide) includes a listing of the C67x floating-point instructions.]

There are two floating-point data representations on the C67x processor: single-precision (SP) and double-precision (DP). In the single precision format, a value is expressed as

$$-1^s * 2^{(exp-127)} * 1.frac \tag{6.5}$$

where s denotes the sign bit (bit 31), *exp* the exponent bits (bits 23 through 30), and *frac* the fractional or mantissa bits (bits 0 through 22). (See Figure 6-6.)

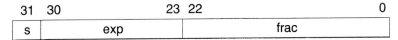

Figure 6-6: Floating point data representation.

Consequently, numbers as big as $3.4*10^{38}$ and as small as $1.175*10^{-38}$ can be processed.

In the double-precision format, more fractional and exponent bits are used as indicated below

$$-1^s * 2^{(exp-1023)} * 1.frac \tag{6.6}$$

where the exponent bits are from bits 20 through 30 and the fractional bits are all the bits of one word and bits 0 through 19 of the other word. (See Figure 6-7.) In this manner, numbers as big as $1.7*10^{308}$ and as small as $2.2 * 10^{-308}$ can be handled.

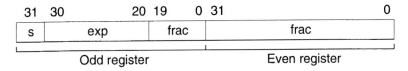

Figure 6-7: Double precision floating point representation.

When using a floating-point processor, all the steps needed to perform floating-point arithmetic are done by the CPU floating-point hardware. For example, consider adding two floating-point numbers represented by

$$a = a_{frac} * 2^{a_{exp}}$$

$$b = b_{frac} * 2^{b_{exp}} \tag{6.7}$$

The floating-point sum c has the following exponent and fractional parts:

$$c = a + b$$

$$= \left(a_{frac} + \left(b_{frac} * 2^{-\left(a_{exp} - b_{exp} \right)} \right) \right) * 2^{a_{exp}} \quad \text{if} \quad a_{exp} \geq b_{exp}$$

$$= \left(\left(a_{frac} * 2^{-\left(b_{exp} - a_{exp} \right)} \right) + b_{frac} \right) * 2^{b_{exp}} \quad \text{if} \quad a_{exp} < b_{exp} \tag{6.8}$$

These parts are computed by the floating-point hardware. This shows that, though possible, it is inefficient to perform floating-point arithmetic on fixed-point processors, since all the operations involved, such as those in Eq.(6.8), must be done in software.

The instructions ending in SP denote single-precision data format and in DP double-precision data format (for example, MPYSP and MPYDP). It should be noted that some of these instructions require additional execute (E) cycles or latencies compared with fixed-point instructions. (See Figure 3-8.) For example, MPYSP requires three delays or NOPs and MPYDP nine delays or NOPs compared with one delay or NOP for fixed-point multiplication MPY.

As illustrated in Figure 6-8, the C62x can support 40-bit and the C67x 64-bit operations by concatenating two registers. Table 6-1 shows a listing of all the C6x datatypes.

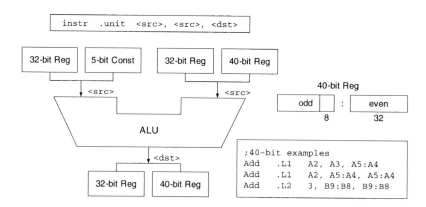

Figure 6-8: 40-bit operations.[†]

Table 6-1: C6x datatypes.[†]

Type	Size	Representation
char, signed char	8 bits	ASCII
unsigned char	8 bits	ASCII
short	16 bits	2's complement
unsigned short	16 bits	binary
int, signed int	32 bits	2's complement
unsigned int	32 bits	binary
long, signed long	40 bits	2's complement
unsigned long	40 bits	binary
enum	32 bits	2's complement
float	32 bits	IEEE 32-bit
double	64 bits	IEEE 64-bit
long double	64 bits	IEEE 64-bit
Pointers	32 bits	binary

6.4 Overflow and Scaling

As stated before, fixed-point processors have a much smaller dynamic range than their floating-point counterparts. Even though the C62 is considered to be a 32-bit device, its multiplier can only multiply 16-bit numbers. It is due to this limitation that the Q-15 representation of numbers is normally considered. The 16-bit multiplier can multiply two Q-15 numbers and produce a 32-bit product. Then the product can be stored in 32 bits or shifted back to 16 bits for storage or further processing.

When multiplying two Q-15 numbers, which are in the range of –1 and 1, it is clear that the resulting number will always be in the same range. However, when two Q-15 numbers are added, the sum may fall outside this range, leading to an overflow. Overflows can cause major problems by generating erroneous results. When using a fixed-point processor, the range of numbers must be closely examined and adjusted to compensate for overflows. The simplest correction method for overflows is scaling.

The idea of scaling can be applied to most filtering and transform operations, where the input is scaled down for processing and the output is then scaled back up to the original size. An easy way to do scaling is by shifting. Since a right shift of 1 is equivalent to a division by 2, we can scale the input repeatedly by 0.5 until all overflows disappear. The output can then be rescaled back to the total scaling amount.

As far as FIR and IIR filters are concerned, it is possible to scale coefficients to avoid overflows. Let us consider the output of a filter $y[n] = \sum_{k=0}^{N-1} h[k] * x[n-k]$, where the h's denote coefficients or unit sample response terms and the x's input samples. In case of IIR filters, for a large enough N, the terms of the unit sample response become so small that they can be ignored. Let us suppose that x's are in Q-15 format (i.e., $|x[n-k]| \leq 1$). Therefore, we can write $|y[n]| \leq \sum_{k=0}^{N-1} |h[k]|$. This means that, to ensure no output overflow (i.e., $|y[n]| \leq 1$), the condition $\sum_{k=0}^{N-1} |h[k]| \leq 1$ must be satisfied. This condition can be satisfied by repeatedly scaling (dividing by 2) coefficients or unit sample response terms.

The C62 provides a saturation flag bit, which is bit 9 of the CSR register. To cope with addition overflows, the saturated add instruction SADD can be used to see whether the saturation bit SAT is set to unity, indicating an overflow. Assuming Q-15 format values, the following function can be used to check the status of the SAT bit after using the _sadd() intrinsic:

```
short safe_add(short A, short B,int *status)
{
    int X,Y,result, SAT_BIT;
    X = A << 16;
    Y = B << 16;
    result = _sadd(X,Y);
    SAT_BIT=(CSR & 0x00000200) >> 9;
    if(SAT_BIT==1){
            //Overflow Occured
            CSR = CSR & 0xFFFFFDFF;        //Reset Sat Bit
            *status = 1;
    }
    else
            *status = 0;

    return (result >> 16);
}
```

This function adds two 16-bit numbers and reports any occurring overflow. If an overflow occurs, it also clears the SAT bit in the CSR.

6.5 Some Useful Arithmetic Operations

The C6x provides useful instructions, such as SUBC, ABS, CMPLT, and NORM, that facilitate efficient implementation of arithmetic operations not available by the

CPU. There are many such operations, including division, trigonometric operations, and square-root. In this section, we provide a number of techniques for implementing these operations. Clearly, it takes many instructions for such operations to get computed. It should also be noted that, in some applications, it is more efficient to implement special purpose arithmetic operations by using lookup tables.

6.5.1 Division

The floating-point C67x DSP provides a reciprocal instruction RCPSP, which gives a good estimate consisting of the correct exponent part and an accurate fractional part up to the eighth binary position. It is possible to extend the accuracy by using this instruction as the seed point $v[0]$ for the iterative Newton-Raphson algorithm expressed by the equation

$$v[n + 1] = v[n] * (2.0 - x * v[n]), \tag{6.9}$$

where x is the value whose reciprocal is to be found. Accuracy is increased by each iteration of this equation. Hence, on the floating-point C67x DSP, division can be achieved by taking reciprocal of the denominator and then by multiplying the reciprocal with the numerator.

On the fixed-point C6x DSP, however, no reciprocal instruction is available. One way to compute reciprocal on the fixed-point C6x is to use the iterative Newton-Raphson equation. However, representing "2" as part of the equation poses a difficulty when values are in Q15 format. This difficulty can be overcome by representing "2" in Q13 instead of Q15 format, noting that the overall accuracy of the reciprocal is reduced to Q13. It is important to choose an initial seed that will allow convergence in as few steps as possible. The code to implement reciprocal with an initial seed of "1" for three iterations is as follows:

```
    mvk    16384,temp2
    shr    x,2,temp1
    sub    temp2,temp1,recip         ; 1st iteration

    mpy    x,recip,temp1
    shr    temp1,13,temp1
    sub    temp2,temp1,temp1
    mpy    recip,temp1,temp1
    shr    temp1,13,recip            ; 2nd iteration

    mpy    x,recip,temp1
    shr    temp1,13,temp1
    sub    temp2,temp1,temp1
    mpy    recip,temp1,temp1
    shr    temp1,13,recip            ; 3rd iteration
```

Another way to implement division is by using the conditional subtraction instruction SUBC. Table 6-2 shows four possible situations, given a 16-bit positive dividend x and a 16-bit positive divisor y. If a dividend or divisor is negative, the absolute instruction ABS should be used to convert them into positive values. The sign of the quotient, of course, will be the same as the sign of the product $x * y$, which is resolved by using the compare instruction CMPLT. The sign of the remainder will be the same as the sign of the dividend. As shown in Table 6-2, all the situations lead to two cases: integer division and fractional division. If $x > y$, the quotient will be in integer format, and if $x < y$, it will be in Q15 format.

Table 6-2: Division types for different datatypes.

	Datatypes	Division type
$x > y$	integers or fractions	integer
$x < y$	integers or fractions	fractional

To do integer division, the SUBC instruction is repeated 16 times. As illustrated in Figure 6-9, the dividend is shifted until subtracting the divisor no longer gives a negative result. Then, for each subtraction that generates a positive result, the result is shifted and a 1 is placed in LSB. After 16 such subtractions, the quotient appears in the low, and the remainder in the high, portion of the dividend register. Such a binary division is obtained in the same manner as long-division.

Long Division:

```
                    000000000000110
0000000000000101 | 000000000100001      Quotient
                    -101
                    ─────
                    110
                    -101
                    ─────
                    11                   Remainder
```

SUBC Method:

31 HIGH register	LOW register 0	Comment
0000000000000000	0000000000100001	(1) Dividend is loaded into register. The divisor is left-shifted 15 and subtracted from register.
-10	1000000000000000	The subtraction is negative, so discard the result
-10	0111111111011111	and shift the register left one bit.
0000000000000000	0000000001000010	(2) 2nd subtract produces negative answer, so
-10	1000000000000000	discard result and shift register (dividend) left.
-10	0111111110111110	
⋮	⋮	⋮
0000000000000100	0010000000000000	(14) 14th SUBC command. The result is positive. Shift
-10	1000000000000000	result left and replace LSB with 1.
0000000000000001	1010000000000000	
0000000000000011	0100000000000001	(15) Result is again positive. Shift result left and
-10	1000000000000000	replace LSB with 1.
0000000000000000	1100000000000001	
0000000000000001	1000000000000011	(16) Last subtract. negative answer, so discard
-10	1000000000000000	result and shift register left.
	1111111111111101	
0000000000000011	0000000000000110	Answer reached after 16 SUBC instructions.
Remainder	Quotient	

Figure 6-9: SUBC division example 33 by 5.[†]

To do fractional division, the same long-division procedure is used. However, this time the SUBC instruction is repeated only 15 times, due to Q15 format representations of the dividend and divisor. The code below shows the division for the fractional case, x and y are assumed to be in Q15 formats and $x < y$:

```
             .global    _divfra
             .title     "_divfra.sa"
_divfra:     .proc A4, B4, B3  ; x<y
             .reg   x, y, count, prod, quot, sign
```

```
            .global    _divfra
            .title     "_divfra.sa"
_divfra:    .proc      A4, B4, B3     ; x<y
            .reg       x, y, count, prod, quot, sign

            MV         A4, x
            MV         B4, y
            MVK        15, count
            ZERO       sign

            MPY        x, y, prod
            CMPLT      prod, 0x0000, sign   ;find quotient sign
            ABS        x, x                 ;make x and y positive
            ABS        y, y
            SHL        x, 15, x             ;not required for integer division
            SH         y, 15, y

loop:       .trip      15                   ;16 for integer division
            SUBC       x, y, x
[count]     SUB        count, 1, count
[count]     B          loop

            MV         x, quot
[sign]      NEG        quot, quot           ;incorporate quotient sign
            MV         quot, A4

            .endproc A4, B3

            B          B3
            NOP        5
```

6.5.2 Sine and Cosine

Trigonometric functions such as sine and cosine can be approximated by using the Taylor series expansion. For sine, we can write the expansion as:

$$sin(x) = x - \frac{x^3}{3!} + \frac{x^5}{5!} - \frac{x^7}{7!} + \frac{x^9}{9!} + \text{higher order} \qquad (6.10)$$

for the first five terms. Adding higher order terms leads to more precision. For implementation purposes, this expansion can be rewritten as follows:

$$sin(x) \cong x * \left(1 - \frac{x^2}{2*3}\left(1 - \frac{x^2}{4*5}\left(1 - \frac{x^2}{6*7}\left(1 - \frac{x^2}{8*9}\right)\right)\right)\right) \qquad (6.11)$$

Similarly, for cosine, we can write:

$$cos(x) \cong 1 - \frac{x^2}{2} + \frac{x^4}{4!} - \frac{x^6}{6!} + \frac{x^8}{8!} = 1 - \frac{x^2}{2}\left(1 - \frac{x^2}{3*4}\left(1 - \frac{x^2}{5*6}\left(1 - \frac{x^2}{7*8}\right)\right)\right) \qquad (6.12)$$

Furthermore, to generate sine and cosine waves, the following recursive formulas can be used:

$$sin\ nx = 2\cos x * sin(n-1)x - sin(n-2)x$$
$$cos\ nx = 2\cos x * cos(n-1)x - cos(n-2)x$$

(6.13)

6.5.3 Square-Root

Square-root $sqrt(y)$ can be approximated by the following Taylor series expansion considering that $y^{0.5} = (x+1)^{0.5}$:

$$sqrt(y) \cong 1 + \frac{x}{2} - \frac{x^2}{8} + \frac{x^3}{16} - \frac{5x^4}{128} + \frac{7x^5}{256}$$

$$= 1 + \frac{x}{2} - 0.5\left(\frac{x}{2}\right)^2 + 0.5\left(\frac{x}{2}\right)^3 - 0.625\left(\frac{x}{2}\right)^4 + 0.875\left(\frac{x}{2}\right)^5$$

(6.14)

Here, it is assumed that x is in Q15 format. In this equation, the estimation error would be small for x values near unity. Hence, to improve accuracy in applications where the range of x is known, x can be scaled by a^2 to bring it close to 1 (i.e., $sqrt(a^2x)$ where $a^2x \cong 1$). The result should then be scaled back by $1/a$.

It is also possible to compute square-root by using the following recursive equation:

$$v[n+1] = v[n] * (1.5 - (x/2) * v[n] * v[n])$$

(6.15)

6.5.4 Lookup Table

A lookup table approach can be adopted to achieve function computation. An example of a lookup table is given next to show how this approach works.

```
shr    x,5,index
ldh    *+p_arctan[index],arctan
```

In this example, the arctangent function $arctan(x)$ is computed based on a previously stored table of length 1024. Since $arctan(-x) = \pi/2 - arctan(x)$, the table only needs to include the entries for positive x values. Arctangent values vary from $-\pi/2$ to $\pi/2$ for x values from -1 to 1.

Bibliography

[1] J. Proakis and D. Manolakis, *Digital Signal Processing: Principles, Algorithms, and Applications*, Prentice-Hall, 1996.

Lab 3: Integer Arithmetic

Implementing algorithms on a fixed-point DSP requires that the range of numbers be closely examined in order to make necessary adjustments to avoid overflows. The simplest approach to correct for overflow is by scaling the input. This lab demonstrates the scaling approach to correct for overflows.

L3.1 Overflow Handling

An overflow occurs when the result of an operation is too large or too small for the CPU to handle. In a 16-bit system, when manipulating integer numbers, they must remain in the range of −32768 to 32767. Otherwise, any operation resulting in a number smaller than −32768 or larger than 32767 will cause overflow. For example, when 32767 is multiplied by 2, we get 65534, which is beyond the representation limit of a 16-bit system.

Consider the following program:

```
#include <stdio.h>

#define SIZE 16

short SIGNAL[SIZE] = {
    11474, 21204, 27709, 29999, 27727, 21238, 11519, 47,
    -11430, -21170, -27691, -29999, -27746, -21272, -11563, -95
    };  // Original data

short NEWSIGNAL[SIZE]; // Data after multiplication

main()
{
    int i;

    for(i = 0; i < SIZE; i++ )
    {
        NEWSIGNAL[i] = SIGNAL[i] * 2; // multiply by 2
    }
}
```

In this program, the array SIGNAL contains samples of a sinusoidal signal. These sample values are multiplied by 2, and the results are placed into the array NEWSIGNAL. Let us examine whether any overflow is caused by these multiplications. In order to monitor the values, it is possible to use either the **Watch Window** or **View Memory** feature of CCS. Let us use the **View Memory** feature by choosing **View →** **Memory** from the menu bar. As a result, the dialog box as shown in Figure 6-10 will appear. In the dialog box, enter SIGNAL in the **Address** field, select **16-Bit Signed Int** from the pop-down list of the **Format** field, and then click **OK**. A memory window displaying the values of SIGNAL will appear, as shown in Figure 6-11(a). Repeat these steps to see the values of NEWSIGNAL, as shown in Figure 6-11(b). From Figure 6-11, it can be seen that the array NEWSIGNAL includes wrong values due to overflows. For example, -23128 is indicated to be the result of the multiplication of 21204 by 2, which is incorrect.

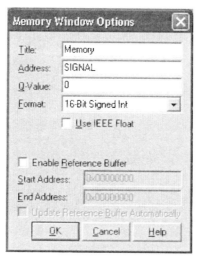

Figure 6-10: Memory Window Options dialog box.

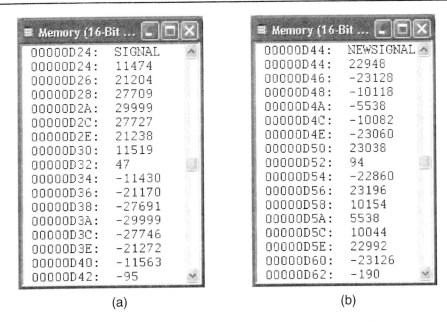

(a) (b)

Figure 6-11: Memory windows showing array values.

As shown in Figure 6-12, the CCS **View Graph** feature can be used to display SIGNAL and NEWSIGNAL. The multiplication of SIGNAL by 2 is expected to generate another sinusoidal signal with twice the amplitude. However, as seen from Figure 6-12(b), NEWSIGNAL is distorted and clipped when the multiplication results are beyond the 16-bit (short datatype) range.

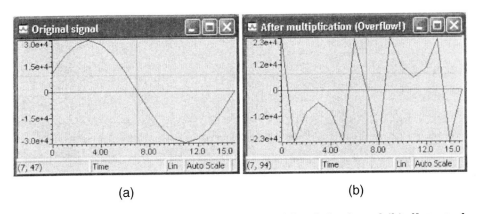

(a) (b)

Figure 6-12: Signal distorted by overflow: (a) original, and (b) distorted.

L3.2 Scaling Approach

Scaling samples is the most widely used approach to overcome the overflow problem. In order to see how scaling works, let's consider a simple multiply/accumulate operation. Suppose there are four constants or coefficients that need to be multiplied with samples of an input analog signal. The worst possible overflow case would be the one where all the multiplicants (C_k's and $x[n]$'s) are 1. For this case, the result $y[n]$ will be 4, given that $y[n] = \sum_{k=1}^{4} C_k * x[n-k]$. Assuming that we have control only over the input, the input samples should be scaled so that the result $y[n]$ will fall in the allowed range. A single right shift reduces the input by one half, and a double shift reduces it further by one quarter. Of course, this leads to less precision, but it is better than getting erroneous results.

A simple method to implement the scaling approach is to create a function that returns the necessary amount of scaling on the input. For any multiply/accumulate type of operations, such as filtering or transform, the worst case is the multiplication and addition of all 1's. Then the required amount of scaling would be dependent on the number of additions in the summation. To examine the worst case, it is required to obtain the required number of scaling so that all overflows disappear. This can be achieved by writing a function to compute the required number of scalings or shiftings of input samples. For the example in this lab, such a function is as follows and is named getNumberOfScaling():

```
#include <stdio.h>
#include <c6x.h>
#define SIZE 16

float Coeff[SIZE] = {0, 0.8311, -0.2977, 0.4961, 0.6488, -0.3401,
        -0.0341, -0.2336, -0.3801, -0.3984, -0.2568, 0.4884,
        0.1113, 0.2495, 0.9999, -0.4088}; /* coefficient */

short safe_add(short A, short B, int *status);
void rescale(short g[]);

void main()
{
    int n;
    n = GetNumberOfScaling(Coeff);

    switch (n)
    {
    case 0: { printf("No scaling is required.\n"); break; }
    case 1: { printf("1 scaling is required to avoid overflow.\n"); break; }
```

```
            default: { printf("%d scalings are required to avoid overflow.\n", n); }
        }
}

int GetNumberOfScaling(float *Coeff)
{
        short sum, g[SIZE];
        int i,bOverFlow, numberOfScaling;

        // Convert to Q-15, good approximate
        for(i=0;i<SIZE;i++)
        {
            g[i]=0x7fff*Coeff[i];
        }
        numberOfScaling = 0;
start:
        sum = 0;
        // Add all values to see if OVERFLOW occurs
        for(i=0;i<SIZE;i++)
        {
            sum = safe_add(sum,g[i],&bOverFlow);
            if(bOverFlow == 1) // Overflow occurred.
            {    rescale(g);
                 numberOfScaling++;
                 printf("Overflow occurred at summation %d\n", i+1);
                 goto start;
            }
        }
        return numberOfScaling;
}

void rescale(short g[])
{
        int k, temp;
        //Rescale Input since it Overflows
        for( k = 0; k < SIZE; k++ )
        {
            temp = (0x4000 * g[k]) << 1;  // Half it
            g[k] = temp >> 16;
        }
}

short safe_add(short A, short B, int *status)
{
        int AA,BB,result,SAT_BIT;
        AA = A << 16;
        BB = B << 16;
        result = _sadd(AA,BB);
        SAT_BIT = (CSR & 0x00000200) >> 9;
        if( SAT_BIT == 1 )        // Overflow Occured
        {
            CSR = CSR & 0xFFFFFDFF; // Reset Sat Bit
            *status = 1;
        }
        else
        {
```

```
        *status = 0;
    }
    return (result >> 16);
}
```

The function GetNumerOfScaling() first produces a good approximation to Q-15 format by multiplying the input float values by 0x7FFF (effectively scaling by 2^{15}). The summation is then obtained by using the function safe_add(), which sets the saturation bit of CSR if an overflow occurs. The value of the CSR register is accessible with the pre-defined variable CSR in the header file *c6x.h*. In order to get bit 9, or the saturation bit, a bitwise AND operation is carried out between CSR and 0x00000200, then the result is right-shifted by 9 bits. The overflow status is checked after every call to safe_add(). If it is 1, indicating an overflow, the function rescale() is called to scale down the input. The number of scalings is also counted. After scaling the input, the summation is repeated. If another overflow occurs, the input sample is scaled down further. This process is continued until no overflow occurs. The final number of scalings is then returned. Care must be taken not to scale the input too many times; otherwise, the input signal gets buried in quantization noise.

It should be noted that, in addition to scaling the input, it is also possible to scale the coefficients or constants in a summation (such as filter coefficients or FFT twiddle factors) to force the outcome to stay within the dynamic range. Depending on the values of constants or coefficients, it may not be necessary to do the maximum shift for each value. As far as the preceding program is concerned, it can be seen that an overflow occurs at the fourth summation, and one scaling is required to avoid it. The execution result is displayed in Figure 6-13, and Table 6-3 shows the sum of the coefficients C_k's. Notice that in the worst case, the inputs are all 1's, so the sum of the C_k's overflows at the fourth summation, which is highlighted in the table. If the coefficients are scaled down by one-half, this is equivalent to scaling down the input samples by one-half, the overflows disappear.

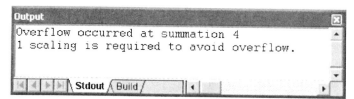

Figure 6-13: Overflow program execution result.

Table 6-3: Scaling example.

C_k	$\sum C_k$	$\dfrac{C_k}{2}$	$\sum \dfrac{C_k}{2}$
0	0	0	0
0.8311	0.8311	0.41555	0.41555
−0.2977	0.5334	−0.14885	0.2667
0.4961	1.0295	0.24805	0.51475
0.6488	1.6783	0.3244	0.83915
−0.3401	1.3382	−0.17005	0.6691
−0.0341	1.3041	−0.01705	0.65205
−0.2336	1.0705	−0.1168	0.53525
−0.3801	0.6904	−0.19005	0.3452
−0.3984	0.292	−0.1992	0.146
−0.2568	0.0352	−0.1284	0.0176
0.4884	0.5236	0.2442	0.2618
−0.1113	0.4123	−0.05565	0.20615
0.2495	0.6618	0.12475	0.3309
0.9999	1.6617	0.49995	0.83085
−0.4088	1.2529	−0.2044	0.62645

7

Code Optimization

Four relatively simple modifications of assembly code can be done to generate a more efficient code. These modifications make use of the available C6x resources such as multiple buses, functional units, pipelined CPU, and memory organization. They include (a) using parallel instructions, (b) eliminating delays or NOPs, (c) unrolling loops, and (d) using word-wide data.

Wherever possible, parallel instructions should be used to make maximum use of idle functional units. It should be noted that, whenever the order in which instructions appear is important, care must be taken not to have any dependency in the operands of the instructions within a parallel instruction.

It may become necessary to have cross paths when making instructions parallel. There are two types of cross paths: data and address. As illustrated in Figure 7-1(a), in data cross paths, one source part of an instruction on the A or B side comes from the other side. A cross path is indicated by x as part of functional unit assignment. The destination is determined by the unit index 1 or 2. As an example, we might have:

```
MPY  .M1x A2,B3,A4
MPY  .M2x A2,B3,B4
```

In address cross paths, a .D unit gets its data address from the address bus on the other side. There are two address buses: DA1 and DA2, also known as T1 and T2, respectively. Figure 7-1(b) illustrates an example where a load and a store are done in parallel via the address cross paths.

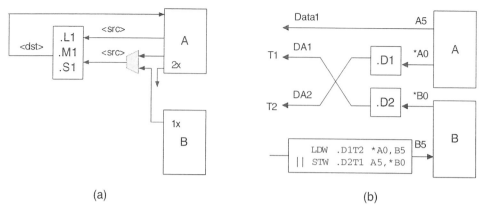

(a)   (b)

Figure 7-1: (a) Data cross-path, and (b) address cross-path.[†]

Wherever possible, branches should be placed five places ahead of where they are intended to appear. This would create a delayed branch, minimizing the number of NOPs. This approach should also be applied to load and multiply instructions that involve four delays and one delay, respectively. If the code size is of no concern, loops should be repeated or copied. By copying or unrolling a loop, fewer clock cycles would be needed, primarily due to deleting branches. Figure 7-2 shows the optimized version of the dot-product loop incorporating the preceding steps.

```
Loop:
            LDH    .D1    *A8++,A2      ;load input 1 into A2
      ||    LDH    .D2    *B9++,B3      ;load input 2 into B3
   [B0]     SUB    .L2    B0,1,B0       ;decrement counter
   [B0]     B      .S1    Loop          ;branch to Loop
            NOP           2             ;5 latency slots required
            MPY    .M1X   A2,B3,A4      ;A4=A2*B3, crosspath
            NOP
            ADD    .L1    A4,A6,A6      ;A6 += A4
```

Figure 7-2: Optimized dot-product example.

Considering that there exists a delay associated with getting information from off-chip memory, program codes should be run from the on-chip RAM whenever possible. In situations where program codes would not fit into the on-chip RAM, faster execution can be achieved by placing the most time-consuming routine or function in the on-chip memory. The C6x has a cache feature which can be enabled to turn the program RAM into cache memory. This is done by setting the program cache control (PCC) bits of the CSR to 010. For repetitive operations or loops, it is recommended that this feature is enabled, since there is then a good chance the cache will contain the needed fetch packet and the EMIF will be unused, speeding up code execution. Figure 7-3 shows the code for enabling the cache feature. The instruction CLR and SET are used to clear and set bits from the second argument position to the third argument position. For more detailed operation of cache and its options, refer to the *CPU Reference Guide* [1].

```
        .def    _enable_cache

_enable_cache:
        b       .s2     B3
        mvc     .s2     CSR, B0
        clr     .s2     B0, 5, 7, B0
        set     .s2     B0, 6, 6, B0
        mcv     .s2     B0, CSR
        nop
```

Figure 7-3: Enabling cache feature.

7.1 Word-Wide Optimization

If data are in halfwords (16 bits), it is possible to perform two loads in one instruction, since the CPU registers are 32 bits wide. In other words, as shown in Figure 7-4, one data can get loaded into the lower part of a register and another one into the upper part.

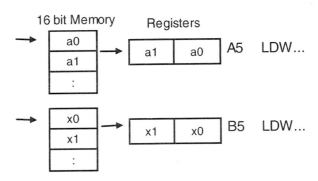

Figure 7-4: Use of LDW to load data.

This way, to do multiplication, two multiplication instructions MPY and MPYH, should be used, one taking care of the lower part and the other of the upper part, as shown in Figure 7-5. Note that A5 and B5 appear as arguments in both MPY and MPYH instructions. This does not pose any conflict, since, on the C6x, up to four reads of a register in one cycle are allowed.

143

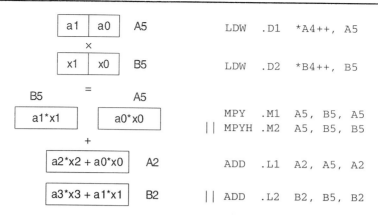

Figure 7-5: Word-wide optimization.†

Figure 7-6 provides the word-wide optimized version of the dot-product function DotP(). When the looping is finished, register A2 would contain the sum of even terms and register B2 the sum of odd terms. To obtain the total sum, these registers are added outside the loop.

```
        .def    DotP

;A4 = &a, B4 = &x, A6 = 20, B3 = return address

DotP:   zero    A2                      ;A2=0
   ||   zero    B2                      ;B2=0
        mv      A6,B0                   ;set B0 to argument passed in A6
loop:
        ldw     .d1     *A4++,A5        ;input word
   ||   ldw     .d2     *B4++,B5        ;input word
  [B0]  sub     .l2     B0,1,B0         ;decrement loop counter
  [B0]  b       .s1     loop            ;branch to loop (5 delay slots filled below)
        nop             2
        mpy     .m1     A5,B5,A5        ;A5=A5(low)*B5(low)
   ||   mpyh    .m2     A5,B5,B5        ;B5=A5(high)*B5(high)
        nop
        add     .l1     A2,A5,A2        ;A2 += A5
   ||   add     .l2     B2,B5,B2        ;B2 += B5

rtn:    b       .s2     B3              ;branch back to calling address
        add     .l1x    A2,B2,A4        ;A4 = A2 + B2 return value
        nop             4
```

Figure 7-6: Word-wide optimized version of dot product code.

Out of the preceding modifications, it is possible to do the last one, word-wide optimization, in C. This demands using an appropriate datatype in C. Figure 7-7 shows the word-wide optimized C code by using the _mpy() and _mpyh() intrinsics.

```
//Prototype
short DotP(int *m, int *n, short count);

//Declarations
short a[40] = {40,39,...1};
short x[40] = {1,2,...40};
short y = 0;
main()
{
        y = DotP((int *)a, (int *)x, 20);

}

short DotP(int *m, int *n, short count)
{
        short i;
        short productl;
        short producth;
        short suml = 0;
        short sumh = 0;

        for(i=0, i<count; i++)
        {
                productl = _mpy(m[i],n[i]);
                producth = _mpyh(m[i],n[i]);
                suml += productl;
                sumh += producth;
        }
        suml += sumh;
        return(suml);
}
```

Figure 7-7: Word-wide optimized code in C.

7.2 Mixing C and Assembly

To mix C and assembly, it is necessary to know the register convention used by the compiler to pass arguments. This convention is illustrated in Figure 7-8. DP, the base pointer, points to the beginning of the .bss section, containing all global and static variables. SP, the stack pointer, points to local variables. The stack grows from higher memory to lower memory, as indicated in Figure 7-8. The space between even registers (odd registers) is used when passing 40-bit or 64-bit values.

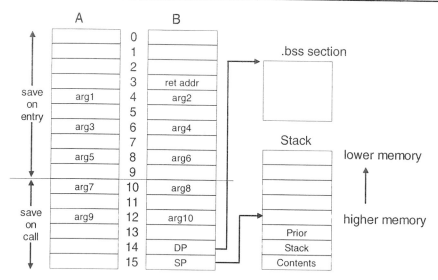

Figure 7-8: Passing arguments convention.†

7.3 Software Pipelining

Software pipelining is a technique for writing highly efficient assembly loop codes on the C6x processor. Using this technique, all functional units on the processor are fully utilized within one cycle. However, to write hand-coded software pipelined assembly code, a fair amount of coding effort is required, due to the complexity and number of steps involved in writing such code. In particular, for complex algorithms encountered in many communications, and signal/image processing applications, hand-coded software pipelining considerably increases coding time. The C compiler at the optimization levels 2 and 3 (−o2 and −o3) performs software pipelining to some degree. (See Figure 4-1.) Compared with linear assembly, the increase in code efficiency when writing hand-coded software pipelining is relatively slight.

7.3.1 Linear Assembly

Linear assembly is a coding scheme that allows one to write efficient codes (compared with C) with less coding effort (compared with hand-coded software pipelined assembly). The assembly optimizer is the software tool that parallelizes linear assembly code across the eight functional units. It attempts to achieve a good compromise between code efficiency and coding effort.

In a linear assembly code, it is not required to specify any functional units, registers, and NOP's. Figure 7-9 shows the linear assembly code version of the dot-product function. The directives .proc and .endproc define the beginning and end, respectively, of the linear assembly procedure. The symbolic names p_m, p_n, m, n, count, prod, and sum are defined by the .reg directive. The names p_m, p_n, and count are associated with the registers A4, B4, and A6 by using the assignment MV instruction.

Table 7-1: Linear assembly directives.[†]

Directive	Description	Restrictions
.call	Calls a function	Valid only within procedures
.cproc	Start a C/C++ callable procedure	Must use with .endproc
.endproc	End a C/C++ callable procedure	Must use with .cproc
.endproc	End a procedure	Must use with .proc; cannot use variables in the register parameter
.mdep	Indicates a memory dependence	Valid only within procedures
.mptr	Avoid memory bank conflicts	Valid only within procedures; can use variables in the register parameter
.no_mdep	No memory aliases in the function	Valid only within procedures
.proc	Start a procedure	Must use with .endproc; cannot use variables in the register parameter
.reg	Declare variables	Valid only within procedures
.reserve	Reserve register use	
.return	Return value to procedure	Valid only within .cproc procedures
.trip	Specify trip count value	Valid only within procedures

```
            .title    "dotp.sa"
            .def      dotp
            .sect     "code"

dotp:       .proc     A4, B4, A6, B3
            .reg      p_m, m, p_n, n, prod, sum, count

            mv    A4, p_m          ;p_m now has the address of m
            mv    B4, p_n          ;p_n now has the address of n
            mv    A6, count        ;count = the number of iterations
            mvk   0, sum           ;sum=0

loop:       .trip     40           ;minimum 40 iterations through loop
            ldh   *p_m++, m        ;load element of m, postincrement pointer
            ldh   *p_n++, n        ;load element of n, postincrement pointer
            mpy   m, n, prod       ;prod=m*n
```

```
        add    prod, sum, sum    ;sum += prod
[count] sub    count, 1, count   ;decrement counter
[count] b      loop              ;branch back to loop
        mv     sum, A4           ;store result in return register A4

        .endproc A4, B3
```

Figure 7-9: Linear assembly code for dot product example.

As per the register convention, the arguments are passed into and out of the procedure via registers A4, B4, A6, and B3. A4 is used to pass the address of m (arg1), B4 the address of n (arg2), and A6 the address of sum (arg3). Register B3, referred to as a preserved register, is passed in and out with no modification. This is done to prevent it from being used by the procedure. Here, this register is used to contain the return address reached by the branch instruction outside of the procedure. Preserved registers must be specified in both input and output arguments while not being used within the procedure. Table 7-1 provides a list of linear assembly directives.

If the number of iterations is known, a .trip directive should be used for the assembler optimizer to generate the pipelined code. For n iterations of a loop, in a pipelined code, the loop is repeated n' times, where n' = n − *prolog length* (prolog will be explained later in the chapter). The number of iterations, n', is known as the minimum trip count. If .trip is greater than or equal to n', only the pipelined code is created. Otherwise, both the pipelined and the non-pipelined code are created. If .trip is not specified, only the non-pipelined code is created. In C, the function _n_assert() is used to provide the same information as .trip.

To further optimize a linear assembly code, partitioning information can be added. Such information consist of the assignment of data paths to instructions.

7.3.2 Hand-Coded Software Pipelining

First let us review the pipeline concept. Figures 7-10(b) and 7-10(c) show a non-pipelined and a pipelined version of the loop code shown in Figure 7-10(a). As can be seen from this figure, the functional units in the non-pipelined version are not fully utilized, leading to more cycles compared with the pipelined version. There are three stages to a pipelined code, named prolog, loop kernel, and epilog. Prolog corresponds to instructions that are needed to build up a loop kernel or loop cycle, and epilog to instructions that are needed to complete all loop iterations. When a loop kernel is established, the entire loop is done in one cycle via one parallel instruction using the maximum number of functional units. This parallelism is what causes a reduction in the number of cycles.

```
loop:      ldh
      ||   ldh
           mpy
           add
```

(a)

cycle\unit	.D1	.D2	.M1	.M2	.L1	.L2	.S1	.S2
1	ldh	ldh						
2			mpy					
3					add			
4	ldh	ldh						
5			mpy					
6					add			
7	ldh	ldh						
8			mpy					
9					add			

(b)

	cycle\unit	.D1	.D2	.M1	.L1
Prolog	1	ldh	ldh		
loop buildup	2	ldh	ldh	mpy	
Loop Kernel	3	ldh	ldh	mpy	add
	4	ldh	ldh	mpy	add
	5	ldh	ldh	mpy	add
Epilog	6			mpy	add
Completing final operations	7				add

(c)

Figure 7-10: (a) A loop example, (b) non-pipelined code, and (c) pipelined code.†

Three steps are needed to produce a hand-coded software pipelined code from a linear assembly loop code: (a) drawing a dependency graph, (b) setting up a scheduling table, and (c) deriving the pipelined code from the scheduling table.

In a dependency graph (see Figure 7-11 for the terminology), the nodes denote instructions and symbolic variable names. The paths show the flow of data and are annotated with the latencies of their parent nodes. To draw a dependency graph for the loop part of the dot-product code, we start by drawing nodes for the instructions and symbolic variable names.

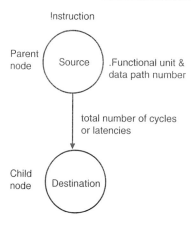

Figure 7-11: Dependency graph terminology.

After the basic dependency graph is drawn, a functional unit is assigned to each node or instruction. Then, a line is drawn to split the workload between the A- and B-side data paths as equally as possible. It is apparent that one load should be done on each side, so this provides a good starting point. From there, the rest of the instructions need to be assigned in such a way that the workload is equally divided between the A- and B-side functional units. The dependency graph for the dot-product example is shown in Figure 7-12.

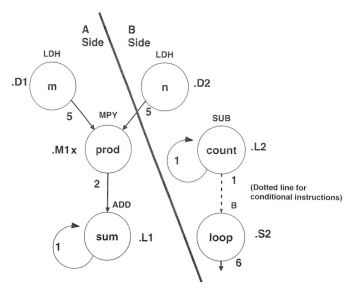

Figure 7-12: Dot-product dependency graph.[†]

The next step for handwriting a pipelined code is to set up a scheduling table. To do so, the longest path must be identified in order to determine how long the table should be. Counting the latencies of each side, we see that the longest path is 8. This means that 7 prolog columns are required before entering the loop kernel. Thus, as shown in Table 7-1, the scheduling table consists of 15 columns (7 for prolog, 1 for loop kernel, 7 for epilog) and eight rows (one row for each functional unit). Epilog and prolog are of the same length.

The scheduling is started by placing the load instructions in parallel in cycle 1. These instructions are repeated at every cycle thereafter. The multiply instruction must appear five cycles after the loads (1 cycle for loads + 4 load delays), so it is scheduled into slot or cycle 6. The addition must appear two cycles after the multiply (1 cycle for multiply + 1 multiply delay), requiring it to be placed in slot or cycle 8, which is the loop kernel part of the code. The branch instruction is scheduled in slot or cycle 3 by reverse counting 5 cycles back from the loop kernel. The subtraction must occur before the branch, so it is scheduled in slot or cycle 2. The completed scheduling table appears in Table 7-2.

Table 7-2: Dot-product scheduling table.†

Cycle Unit		PROLOG						LOOP		EPILOG					
	1	2	3	4	5	6	7	8	9	10	11	12	13	14	15
.D1	①LDH	LDH	LDH	LDH	LDH	LDH	LDH	LDH							
.D2	④LDH	LDH	LDH	LDH	LDH	LDH	LDH	LDH							
.L1								③ADD	ADD	ADD	ADD	ADD	ADD	ADD	ADD
.L2		⑥SUB	SUB	SUB	SUB	SUB	SUB	SUB							
.S1															
.S2			⑤ B	B	B	B	B	B							
.M1							②MPY	MPY	MPY	MPY	MPY	MPY	MPY	MPY	
.M2															

Next, the code is handwritten directly from the scheduling table as 7 prolog parallel instructions, 40 − 7=33 loop kernel parallel instructions, and 7 epilog parallel instructions. This hand-coded software pipelined code is shown in Figure 7-13. It can be seen that this pipelined code requires only 47 cycles to perform the dot-product 40 times. Note that, as shown in Figure 7-14, it is possible to eliminate the epilog instructions by performing the loop kernel instruction 40 instead of 33 times, leading to a lower code size and a higher number of loads.

```
cycle 1:                                    cycle 8 to n: Single-cycle loop
            ldh   .D1   *A1++,A2
     ||     ldh   .D1   *B1++,B2            loop:      ldh   .D1   *A1++,A2
                                                ||     ldh   .D1   *B1++,B2
cycle 2:                                         || [B0] sub  .L2   B0,1,B0
            ldh   .D1   *A1++,A2                 || [B0] B    .S2   loop
     ||     ldh   .D1   *B1++,B2                 ||     mpy   .M1x  A2,B2,A3
     || [B0] sub  .L2   B0,1,B0                  ||     add   .L1   A4,A3,A4

cycle 3,4 and 5:                            cycle n+1 to n+5:
            ldh   .D1   *A1++,A2                        mpy   .M1x  A2,B2,A3
     ||     ldh   .D1   *B1++,B2                 ||     add   .L1   A4,A3,A4
     || [B0] sub  .L2   B0,1,B0
     || [B0] B    .S2   loop               cycle n+6 to n+7:
                                                       add   .L1   A4,A3,A4
cycle 6 and 7:
            ldh   .D1   *A1++,A2
     ||     ldh   .D1   *B1++,B2
     || [B0] sub  .L2   B0,1,B0
     || [B0] B    .S2   loop
     ||     mpy   .M1x  A2,B2,A3
```

Figure 7-13: Hand-coded software pipelined dot-product code.

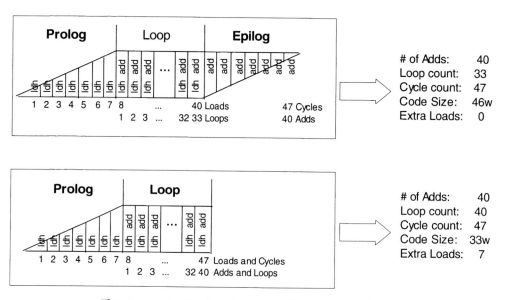

Figure 7-14: Elimination of epilog instructions.†

Figures 7-15(b) and 7-15(c) show the dependency graph and the scheduling table, respectively, of the word-wide optimized dot-product code appearing in Figure 7-15(a). The corresponding hand-coded pipelined code is shown in Figure 7-16. This time, 28 cycles are required: 7 prolog instructions, 40/(2 word datatype) = 20 loop kernel instructions, and one extra add to sum the even and odd parts. Table 7-3 provides the number of cycles for different optimizations of the dot-product example discussed throughout the book. The interested reader is referred to the *TI TMS320C6x Programmer's Guide* [2] for more details on how to handwrite software pipelined assembly code.

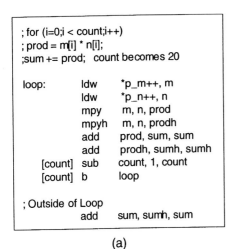

(a)

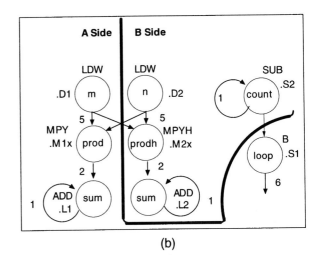

(b)

		PROLOG							LOOP
Unit \ Cycle	1	2	3	4	5	6	7	8
.D1	① ldw m	ldw	ldw	ldw	ldw	ldw	ldw	ldw
.D2	④ ldw n	ldw	ldw	ldw	ldw	ldw	ldw	ldw
.L1								③ add
.L2								⑥ add
.S1			⑧ B	B	B	B	B	B
.S2		⑦ sub	sub	sub	sub	sub	sub	sub
.M1						② mpy	mpy	mpy
.M2						⑤ mpyh	mpyh	mpyh

(c)

Figure 7-15: (a) Linear assembly dot-product code, (b) corresponding dependency graph, and (c) scheduling table.[†]

```
cycle 1:                                cycle 8 to n+7: Single-cycle loop
            ldw   .D1   *A4++,A5
    ||      ldw   .D1   *B4++,B5        loop:       ldw   .D1   *A4++,A5
                                            ||      ldw   .D1   *B4++,B5
cycle 2:                                    || [B0] sub   .S2   B0,1,B0
            ldw   .D1   *A4++,A5            || [B0] B     .S1   loop
    ||      ldw   .D1   *B4++,B5            ||      mpy   .M1x  A5,B5,A6
    || [B0] sub   .S2   B0,1,B0            ||      mpyh  .M2x  A5,B5,B6
                                            ||      add   .L1   A7,A6,A7
cycle 3,4 and 5:                            ||      add   .L2   B7,B6,B7
            ldw   .D1   *A4++,A5
    ||      ldw   .D1   *B4++,B5
    || [B0] sub   .S2   B0,1,B0
    || [B0] B     .S1   loop

cycle 6 and 7:
            ldw   .D1   *A4++,A5
    ||      ldw   .D1   *B4++,B5
    || [B0] sub   .S2   B0,1,B0
    || [B0] B     .S1   loop
    ||      mpy   .M1x  A5,B5,A6
    ||      mpyh  .M2x  A5,B5,B6
```

Figure 7-16: Hand-coded pipelined code for word-wide dot-product loop.

Table 7-3: Optimization methods cycles.

No optimization	16 cycles * 40 iterations =	640
Parallel optimization	15 cycles * 40 iterations =	600
Filling delay slots	8 cycles * 40 iterations =	320
Word wide optimizations	8 cycles * 20 iterations =	160
Software pipelined -LDH	1 cycle * 40 loops + 7 prolog =	47
Software pipelined -LDW	1 cycle * 20 loops + 7 prolog + 1 epilog =	28

Multicycle loops – Let us now examine an example denoting a weighted vector sum $c = a + r * b$, where a and b indicate two arrays or vectors of size 40 and r a constant or scalar. Figure 7-17 shows the linear assembly code and the corresponding dependency graph to compute c. A problem observed in this dependency graph is that there are more than two loads/stores operations (i.e., the .D1 unit is assigned to two nodes). This, of course, is not possible in a single-cycle loop. Consequently, we must have two instead of one cycle loop. In other words, two parallel instructions are needed to compute the vector sum c per iteration.

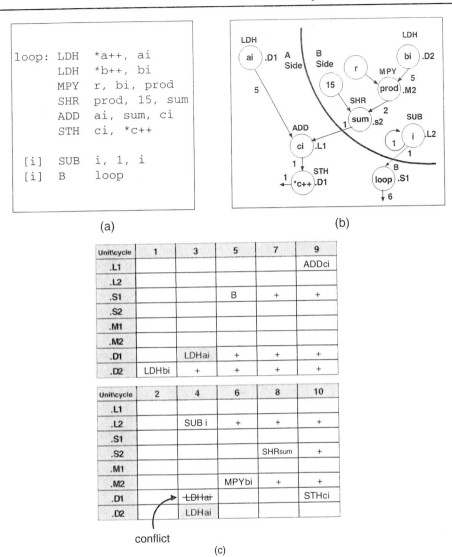

```
loop:  LDH   *a++, ai
       LDH   *b++, bi
       MPY   r, bi, prod
       SHR   prod, 15, sum
       ADD   ai, sum, ci
       STH   ci, *c++

[i]    SUB   i, 1, i
[i]    B     loop
```

(a)

(b)

Unit\cycle	1	3	5	7	9
.L1					ADDci
.L2					
.S1			B	+	+
.S2					
.M1					
.M2					
.D1		LDHai	+	+	+
.D2	LDHbi	+	+	+	+

Unit\cycle	2	4	6	8	10
.L1					
.L2		SUB i	+	+	+
.S1					
.S2				SHRsum	+
.M1					
.M2			MPYbi	+	+
.D1		~~LDHai~~			STHci
.D2		LDHai			

conflict

(c)

Figure 7-17: Multicycle loop: (a) loop code, (b) dependency graph, and (c) scheduling table.[†]

This time, the scheduling table consists of two sets of functional units arranged as shown in Figure 7-17. In this example, the length of the longest path is 10, which corresponds to the load-multiply-shift-add-store path. This means that there should be 10 cycle columns. However, this time the cycle number is set up by alternating between the two sets of functional units. The scheduling is started by entering the instructions for the longest path. The load LDH bi is placed in slot 1. MPY is placed in slot 6, five slots after slot 1 to accommodate for the load latencies. SHR is placed in slot 8, two slots after slot 6 to accommodate for the multiply latency. ADD is placed in slot 9, one slot after slot 8, and STH ci in the last slot 10. The other path is then scheduled. The loading LDH ai for this path must be done 5 slots or cycles before the ADD instruction. This would place LDH ai in slot 4. However, notice that the .D1 unit as part of the second loop cycle has already been used for STH ci and cannot be used at the same time. Hence, this creates a conflict which must be resolved. As indicated by the shaded boxes in Figure 7-17, the conflict is resolved either by using the .D2 unit in the second cycle loop or by using the .D1 unit in the first cycle loop.

7.4 C64x Improvements

This section shows how the additional features of the C64x DSP can be used to further optimize the dot-product example. Figure 7-18(b) shows the C64x version of the dot-product loop kernel for multiplying two 16-bit values. The equivalent C code appears in Figure 7-18(a).

```
main()
{
    y = DotP(a,x,40);
}
int DotP(short *m, short *n, int count)
{
    int i;
    int product;
    int sum = 0;
    for(i=0;i<count;i++)
    {
        product = m[i] * n[i];
        sum += product;
    }
    return(sum);
}
```
(a)

```
;PIPED LOOP KERNEL
LOOP:
        [ A0]  SUB    .L1    A0,1,A0
    ||  [!A0]  ADD    .S1    A6,A5,A5    ;keep running sum
    ||         MPY    .M1X   B4,A4,A6    ;multiply two 16-bit values
    ||  [ B0]  BDEC   .S2    LOOP,B0     ;decrement loop counter and branch if > 0
    ||         LDH    .D1T1  *A3++,A4    ;load 16-bit value
    ||         LDH    .D2T2  *B5++,B4    ;load 16-bit value
```
(b)

Figure 7-18: C64x pipelined code: (a) C, and (b) assembly.[†]

Now, by using the DOTP2 instruction of the C64x, we can perform two 16*16 multiplications, reducing the number of cycles by one-half. This requires accessing two 32-bit values every cycle. As shown in Figure 7-19(a), in C, these can be achieved by using the intrinsic _dotp2() and by casting shorts as integers. The equivalent loop kernel code generated by the compiler is shown in Figure 7-19(b), which is a double-cycle loop containing four 16 * 16 multiplications. The instruction LDW is used to bring in the required 32-bit values.

```
main()
{
    y = DotP((int )a, (int *)x,20);
}

int DotP(int *m, int *n, int count)
{
    int i,
    int product;
    int sum = 0;
    for(i=0;i<count;i++)
    {
        product = _dotp2(m[i], n[i]);
        sum = product + sum;
    }
    return(sum);
}
```

(a)

```
;PIPED LOOP KERNEL
LOOP:
      [!A1] ADD    .L2   B8,B4,B4    ;running sum 0
   ||       DOTP2 .M2X  B7,A6,B8    ;2 16x16 multiplies + add ; prod 0
   ||  [ A0] BDEC   .S1   LOOP, A0    ;decrement loop counter and branch if > 0
   ||       LDW    .D1T1 *+A4(4),A3  ;load a 32-bit value
   ||       LDW    .D2T2 *+B5(4),B6  ;load a 32-bit value
       [ A1] SUB    .L1   A1,1,A1
   ||  [!A1] ADD    .S1   A7,A5,A5    ;running sum1
   ||       DOTP2 .M1X  B6,A3,A7    ;2 16x16 multiplies + add; prod 1
   ||       LDW    .D1T1 *++A4(8), A6   ;load a 32-bit value
   ||       LDW    .D2T2 *++B5(8), B7   ;load a 32-bit value
```

(b)

Figure 7-19: C64x packed datatype code: (a) C, and (b) assembly.[†]

Considering that the C64x can bring in 64-bit data values by using the double-word loading instruction LDDW, the foregoing code can be further improved by performing four 16 * 16 multiplications via two DOTP2 instructions within a single-cycle loop, as shown in Figure 7-20(b). This way the number of operations is reduced by four-fold, since four 16 * 16 multiplications are done per cycle. To do this in C, we need to cast short datatypes as doubles, and to specify which 32 bits of 64-bit data a DOTP2 is supposed to operate on. This is done by using the _lo() and _hi() intrinsics to specify the lower and the upper 32 bits of 64-bit data, respectively. Figure 7-20(a) shows the equivalent C code.

```
int DotP(const short * restrict m, const short * restrict n, int count)
{
    int i;
    int sum = 0;
    const double * restrict m_dbl = (const double *) m;
    const double * restrict n_dbl = (const double *) n;

    count /= 2;         // count is divided by two if using same
                        // main function to call this subroutine

    for(i=0;i<count;i++)
    {
        sum +=  _dotp2(_lo(m_dbl[i]), _lo(n_dbl[i])) +
                _dotp2(_hi(m_dbl[i]), _hi(n_dbl[i]));
    }
    return  sum ;
}
```

(a)

```
;PIPED LOOP KERNEL
LOOP:
    [ B0] SUB     .L2     B0,1,B0       ;decrement running sum counter
||  [!B0] ADD     .S2     B8,B6,B6      ; running sum 0
||  [!B0] ADD     .L1     A7,A6,A6      ; running sum 1
||        DOTP2   .M2X    B4,A4,B8      ; 2 16x16 multiplies + add; prod 0
||        DOTP2   .M1X    B5,A5,A7      ; 2 16x16 multiplies + add; prod 1
||  [A0]  BDEC    .S1     LOOP,A0       ;branch to loop & decrement loop count
||        LDDW    .D1T1   *A3++,A5:A4   ;load a 64-bit value
||        LDDW    .D2T2   *B7++,B5:B4   ;load a 64-bit value
```

(b)

Figure 7-20: C64x double-word packed datatype code: (a) C, and (b) assembly.[†]

Bibliography

[1] Texas Instruments, *TMS320C6000 CPU and Instruction Set Reference Guide,* Literature ID# SPRU 189F, 2000.

[2] Texas Instruments, *TMS320C62x/C67x Programmer's Guide,* Literature ID# SPRU 198B, 1998.

Lab 4: Real-Time Filtering

The purpose of this lab is to design and implement a finite impulse response filter on the C6x. The design of the filter is done by using MATLAB™. Once the design is completed, the filtering code is inserted into the sampling shell program as an ISR to process live signals in real-time.

L4.1 Design of FIR Filter

MATLAB or filter design packages can be used to obtain the coefficients for a desired FIR filter. To have a more realistic simulation, a composite signal may be created and filtered in MATLAB. A composite signal consisting of three sinusoids, as shown in Figure 7-21, can be created by the following MATLAB code:

```
Fs=8e3;
Ts=1/Fs;
Ns=512;

t=[0:Ts:Ts*(Ns-1)];

f1=750;
f2=2500;
f3=3000;

x1=sin(2*pi*f1*t);
x2=sin(2*pi*f2*t);
x3=sin(2*pi*f3*t);

x=x1+x2+x3;
```

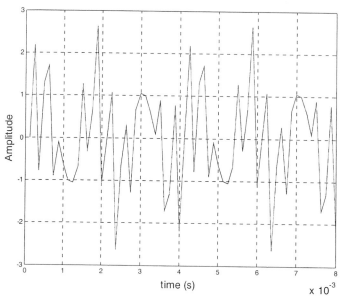

Figure 7-21: Two cycles of composite signal.

The signal frequency content can be plotted by using the MATLAB 'fft' function. Three spikes should be observed, at 750 Hz, 2500 Hz, and 3000 Hz. The frequency leakage observed on the plot is due to windowing caused by the finite observation period. A lowpass filter is designed here to filter out frequencies greater than 750 Hz and retain the lower components. The sampling frequency is chosen to be 8 kHz, which is common in voice processing. The following code is used to get the frequency plot shown in Figure 7-22:

```
X=(abs(fft(x,Ns)));
y=X(1:length(X)/2);
f=[1:1:length(y)];
plot(f*Fs/Ns,y);
grid on;
```

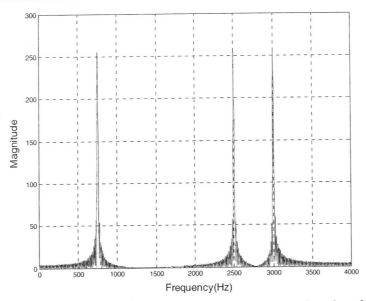

Figure 7-22: Frequency components of composite signal.

To design a FIR filter with passband frequency = 1600 Hz, stopband frequency = 2400 Hz, passband gain = 0.1 dB, stopband attenuation = 20 dB, sampling rate = 8000 Hz, the Parks-McClellan method is used via the 'remez' function of MATLAB [1]. The magnitude and phase response are shown in Figure 7-23, and the coefficients are given in Table 7-3. The MATLAB code is as follows:

```
rp = 0.1;            % Passband ripple
rs = 20;             % Stopband ripple
fs = 8000;           % Sampling frequency
f = [1600 2400];     % Cutoff frequencies
a = [1 0];           % Desired amplitudes
% Compute deviations
dev = [(10^(rp/20)-1)/(10^(rp/20)+1)  10^(-rs/20)];
[n,fo,ao,w] = remezord(f,a,dev,fs);
B = remez(n,fo,ao,w);
A=1;
freqz(B,A);
```

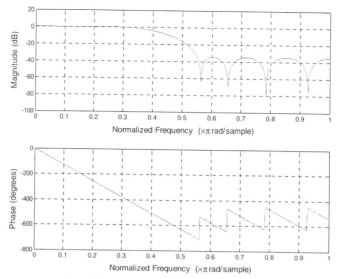

Figure 7-23: Filter magnitude and phase response.

Table 7-3: FIR filter coefficients.

Coefficient	Values	Q-15 Representation
B_0	0.0537	0x06DF
B_1	0.0000	0x0000
B_2	–0.0916	0xF447
B_3	–0.0001	0xFFFD
B_4	0.3131	0x2813
B_5	0.4999	0x3FFC
B_6	0.3131	0x2813
B_7	–0.0001	0xFFFD
B_8	–0.0916	0xF447
B_9	0.0000	0x0000
B_{10}	0.0537	0x06DF

(Note: Do not confuse B coefficients with B registers!)

With these coefficients, the 'filter' function of MATLAB is used to verify that the FIR filter is actually able to filter out the 2.5 kHz and 3 kHz signals. The following MATLAB code allows one to visually inspect the filtering operation:

```
% Figure 7-24
subplot(3,1,1);
va_fft(x,1024,8000);
subplot(3,1,2);
[h,w]=freqz(B,A,512);
plot(w/(2*pi),10*log(abs(h)));
grid on;
```

```
subplot(3,1,3);
y = filter(B,A,x);
va_fft(y,1024,8000);
```

```
function va_fft(x,N,Fs)

X=fft(x,N);
XX=(abs(X));
XXX=XX(1:length(XX)/2);
y=XXX;
f=[1:1:length(y)];
plot(f*Fs/N,y);
grid on;
```

```
% Figure 7-25
n=128
subplot(2,1,1);
plot(t(1:n),x(1:n));
grid on;
xlabel('Time(s)');
ylabel('Amplitude');
title('Original and Filtered Signals');
subplot(2,1,2);
plot(t(1:n),y(1:n));
grid on;
xlabel('Time(s)');
ylabel('Amplitude');
```

Looking at the plots appearing in Figures 7-24 and 7-25, we see that the filter is able to remove the desired frequency components of the composite signal. Observe that the time response has an initial setup time causing the first few data samples to be inaccurate. Now that the filter design is complete, let us consider the implementation of the filter.

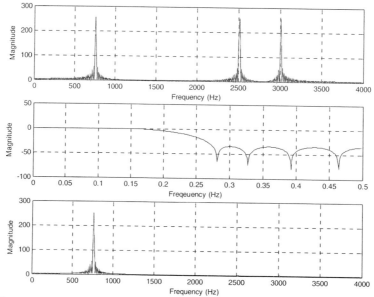

Figure 7-24: Frequency representation of filtering operation.

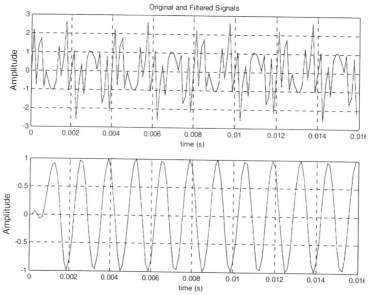

Figure 7-25: Time domain representation of filtering operation.

L4.2 FIR Filter Implementation

An FIR filter can be implemented in C or assembly. The goal of the implementation is to have a minimum cycle time algorithm. This means that to do the filtering as fast as possible in order to achieve the highest sampling frequency (the smallest sampling time interval). Initially, the filter is implemented in C, since this demands the least coding effort. Once a working algorithm in C is obtained, the compiler optimization levels (i.e., –o2, –o3) are activated to reduce the number of cycles. An implementation of the filter is then done in hand-coded assembly, which can be software pipelined for optimum performance. A final implementation of the filter is performed in linear assembly, and the timing results are compared.

The difference equation $y[n] = \sum_{k=0}^{N-1} B_k * x[n-k]$ is implemented to realize the filter. Since the filter is implemented on the DSK, the coding is done by modifying the sampling program in Lab 2, which uses an ISR that is able to receive a sample from the serial port and send it back out without any modification.

When using the C6711 DSK together with the audio daughter card, the received data from McBSP1 is 32-bit wide with the most significant 16 bits coming from the right channel, and the least significant 16 bits coming from the left channel. The FIR filter implementation on the EVM is discussed in Section L4.4.

Considering Q-15 representation here, the MPY instruction is utilized to multiply the lower part of a 32-bit sample (left channel) by a 16-bit coefficient. In order to store the product in 32 bits, it has to be left shifted by one to get rid of the extended sign bit. Now, to export the product to the codec output, it must be right shifted by 16 to place it in the lower 16 bits. Alternatively, the product may be right shifted by 15 without removing the sign bit.

To implement the algorithm in C, the _mpy() intrinsic and the shift operators '<<' and '>>' should be used as follows:

```
result = ( _mpy(sample,coefficient) ) << 1;
result = result >> 16;
```

or

```
result = ( _mpy(sample,coefficient) ) >> 15;
```

Here, result and sample are 32 bits wide, while coefficient is 16 bits wide. The intrinsic _mpy() multiplies the lower 16 bits of the first argument by the lower 16 bits of the second argument. Therefore, the lower 16 bits of sample is used in the multiplication.

For the proper operation of the FIR filter, it is required that the current sample and N-1 previous samples be processed at the same time, where N is the number of coefficients. Hence, the N most current samples have to be stored and updated with each incoming sample. This can be done easily via the following code:

```
void interrupt serialPortRcvISR()
{
    int i, temp, result= 0;
    temp = MCBSP_read(hMcbsp);

    // Update array samples
    for(i=N-1;i>=0;i--)
        samples[i+1] = samples[i];

    samples[0] = temp;

    MCBSP_write(hMcbsp, result);
}
```

Here, as a new sample comes in, each of the previous samples is moved into the next location in the array. As a result, the oldest sample `sample[N]`, is discarded, and the newest sample, `temp`, is put into `sample[0]`.

This approach adds some overhead to the ISR, but for now it is acceptable, since at a sampling frequency of 8 kHz, there is a total of 18,750 cycles are available, [1/(8 kHz /150 MHz) = 18,750], between consecutive samples, considering that the DSK runs at 150 MHz. The total overhead for this manipulation is 358 cycles without any optimization. It should be noted that the proper way of doing this type of filtering is by using circular buffering. The circular buffering approach will be discussed in Lab 5.

Now that the N most current samples are in the array, the filtering operation may get started. All that needs to be done, according to the difference equation, is to multiply each sample by the corresponding coefficient and sum the products. This is achieved by the following code:

```
interrupt void serialPortRcvISR()
{
    int i, temp, result = 0;
    temp = MCBSP_read(hMcbsp);

    // Update array samples
    for( i = N-1 ; i >= 0 ; i-- )
        samples[i+1] = samples[i];

    samples[0] = temp;

    // Filtering
    for( i = 0 ; i <= N ; i++ )
        result += ( _mpy(samples[i], coefficients[i]) ) << 1;
    result = result >> 16;
    MCBSP_write(hMcbsp, result);
}
```

To complete the FIR filter implementation, we need to incorporate the filter coefficients previously designed into the C program. This is accomplished by modifying the sampling program in Lab 2 as follows:

```
    .
    .
#define N 10

// FIR filter coefficients
short coefficients[N+1] = {  0x6DF,  0x0,  0xF447,  0xFFFD,  0x2813,  0x3FFC,
0x2813, 0xFFFD, 0xF447, 0x0, 0x6DF};

int samples[N];
    .
    .
int main()
{
    .
    .

    for(i = 0; i <= N; i++ )
        samples[i]=0;
    .
    .
}
```

The filtering program can now be built and run. Using a function generator and an oscilloscope, it is possible to verify that the filter is working as expected. The output of the function generator should be connected to the line-in jack of the audio daughter card, and the line-out jack of the daughter card to the input of the oscilloscope. As the input frequency is increased, it is seen that the signal attenuation starts at 1.6 kHz and dies out at 2.4 kHz. Note that the nearest sampling frequency should be selected, since the sampling frequencies of the audio daughter card are limited. As a result, the frequencies for passband and stopband are considered to be 1.46 kHz and 2.20 kHz with the sampling frequency of 7324.22 Hz.

Given that a working design is reached, it is time to start the optimization of the filtering algorithm. The first step in optimization is to use the compiler optimizer. The optimizer can be invoked by choosing **Project → Options** from the CCS menu bar. This option will invoke a dialog box, as shown in Figure 7-26. In this dialog box, select **Basic** in the **Category** field, and then choose the desired optimization level from the pull-down list in the **Opt Level** field. Table 7-4 summarizes the timing results for different optimization levels.

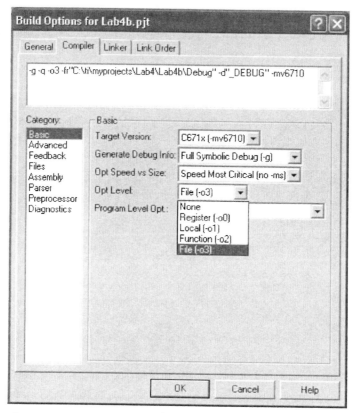

Figure 7-26: Selection of different optimization levels.

Table 7-4: Timing cycles for different builds.

Build Type	Number of Cycles
Compile without optimization	670
Compile with –o0	396
Compile with –o1	327
Compile with –o2/–o3	159

As can be seen from Table 7-4, the number of cycles diminishes as the optimization level is increased. It is important to remember that because the compiler optimizer changes the flow of a program, the debugger may not work in some cases. Therefore, it is advised that one make sure a program works correctly before performing any compiler optimization.

Before doing the linear assembly implementation, the code is written in assembly to see how basic optimization methods such as placing instructions in parallel, filling delay slots, loop unrolling, and word-wide optimization affect the timing cycle of the code.

To perform the operation of multiplying and adding N coefficients, a loop needs be set up. This can be done by using a branch instruction. A counter is required to exit the loop once N iterations have been performed. For this purpose, one of the conditional registers (A1, A2, B0, B1 or B2) is used. No other register allows for conditional testing. Adding [A2] in front of an instruction permits the processor to execute the instruction if the value in A2 does not equal zero. If A2 contains zero, the instruction is skipped, noting that an instruction cycle is still consumed. The .S1 unit may be used to perform the move constant and branch operations. The value in the conditional register A2 decreases by using a subtract instruction. Since the subtract operation should stop if the value drops below zero, this conditional register is mentioned in the SUB instruction to execute it only if the value is not equal to zero. The programmer should remember to add five delay slots for the branch instruction. The code for this loop is as follows:

```
        MVK    .S1    11, A2          ;move 11 into A2 count register
Loop1:
        .
        .
  [A2] SUB   .L1    A2,1,A2          ;decrement counter
  [A2] B     .S1    Loop1            ;branch back to Loop1
        NOP    5
```

We can now start adding instructions to perform the multiplication and accumulation of the values. First, those values that are to be multiplied need to be loaded from their memory locations into the CPU registers. This is done by using load word (LDW) and load half-word (LDH) instructions. Upon executing the load instructions, the pointer is post-incremented so that it is pointed to the next memory location. Once the values have appeared in the registers (four cycles after the load instruction), the MPY instruction is used to multiply them and store the product in another register. Then, the summation is performed by using the ADD instruction. The completed assembly program is as follows:

```
        .global    _fir_simple
        .sect      ".fir_simple"
_fir_simple:
        MV    .S1    A6,A2          ;Count register
        ZERO  .S1    A9             ;Sum register
```

```
loop: LDW    .D1    *A4++,A7    ;Load data from samples
      LDH    .D2    *B4++,B7    ;Load data from coefficients
      NOP    4
      MPY    .M1x   A7,B7,A8    ;A7 is 32 bit sample
                                ;B7 is Q-15 representation coefficient
      NOP
      SHL    A8,1,A8            ;Eliminate sign extension bit
      ADD    .S1    A8,A9,A9    ;Accumulate result
[A2]  SUB    .S1    A2,1,A2     ;Decrement counter
[A2]  B      .S1    loop
      NOP    5

      MV     .S1    A9,A4       ;Move result to return register
      B      .S2    B3          ;Branch back to calling address
      NOP    5
```

The preceding code is a C callable assembly function. To call this function from C, a function declaration must be added as external (extern) without any arguments. The arguments to the function are passed via registers A4, B4 and A6. The return value is stored in A4. Here, the pointers to the arrays are passed in A4 and B4 as the first two arguments and the number of iterations in A6 as the third argument. The return address from the function is stored in B3. Therefore, a final branch to B3 is required to return from the function. For a complete explanation of calling assembly functions from C, see the *TI TMS320C6x Optimizing C Compiler User's Guide* [2].

The directive .sect is used to place the code in the appropriate memory location. Running the code from the external SDRAM memory takes a total of 1578 cycles for the assembly function to complete. To move the code into the internal memory so that it runs faster, the linker command file should be modified by replacing .fir_simple > CE0 with .fir_simple > IRAM. Running the code from the internal memory results in 313 cycles. Notice that only the assembly function is running in the internal program memory; the rest of the ISR is located in the external memory and still runs slow, taking a total of 2938 cycles. It is possible to move the entire ISR into the internal memory to obtain a faster execution. The CODE_SEC-TION pragma can be used for this purpose. By adding the following line to the code, the entire ISR is placed in the internal memory, leading to a total of 754 cycles:

```
#pragma CODE_SECTION(serialPortRcvISR,".isr")
```

To optimize the foregoing function, basic optimization methods, such as placing instructions in parallel, filling delay slots, and loop unrolling, are applied. Examining the code, one sees that some of the instructions can be placed in parallel. Because of operand dependencies, care must be taken not to schedule parallel instructions that use previous operands as their operands. The two initial load instructions are independent and can be made to run in parallel. Looking at the rest of the program, we can see that the operands are dependent on the previous operands; hence, no other instructions are placed in parallel.

To reduce the cycles taken by the NOP instructions, we can use the delay slot filling technique. For example, as the load instructions are executed in parallel, it is possible to schedule the subtraction of the loop counter in place of their NOPs. The branch instruction takes five cycles to execute. It is therefore possible to slide the branch instruction four slots up to get rid of its NOPs. Incorporating these optimizations, we can rewrite the function as follows:

```
        .global   _fir_filled
        .sect     ?.fir_filled?        ;used to load into internal program memory

_fir_filled:
        MV      .S1    A6,A2           ;Count register
        ZERO    .S1    A9              ;Sum register

loop:   LDW     .D1    *A4++,A7        ;Load data from samples
  ||    LDH     .D2    *B4++,B7        ;Load data from coefficients
        NOP
[A2]    SUB     .S1    A2,1,A2         ;Decrement counter
[A2]    B       .S1    loop            ;branch back to loop
        NOP
        MPYHL   .M1x   A7,B7,A8        ;A7 is 32 bit sample,

                                       ;B7 is Q-15 representation coefficient

        NOP
        SHL            A8,1,A8         ;Eliminate sign extension bit
        ADD     .S1    A8,A9,A9        ;Accumulate result

        MV      .S1    A9,A4           ;Move result to return register
        B       .S2    B3              ;Branch back to calling address
        NOP     5
```

By filling delay slots, the number of cycles is reduced. In repetitive loops such as this one, it is seen that the branch instruction takes up extra cycles that can be eliminated. As just mentioned, one method to do this elimination is to fill the delay slots by sliding the branch instruction higher in the execution phase, thus filling the latencies associated with branching. Another method for reducing the latencies is to unroll the loop. However, notice that loop unrolling eliminates only the last latency of the branch. Since, in the preceding delay filled version, the branch latency has no effect on the number of cycles, loop unrolling does not achieve any further improvement in timing.

To perform word-wide optimization, the ISR has to be modified to store a sample into 16 bits rather than 32 bits. This can be simply achieved by using a variable of type short. The following code stores a sample into a short variable, temp, assuming that the result is 32 bits.

```
interrupt void serialPortRcvISR (void)
{
      int i,result = 0;
      short temp;
      temp = MCBSP_read(hMcbsp);          // Takes lower 16 bits only
      //Filtering
      MCBSP_write(hMcbsp, result);
}
```

Using word-wide optimization, one needs to load two consecutive 16 bit values in memory with a single load-word instruction. This way, the input register contains two values, one in the lower and the other in the upper part. The instructions MPYH and MPY can be used to multiply the upper and lower parts, respectively. The following assembly code shows how this is done for the FIR filtering program:

```
      .global      _fir_wordoptimized

_fir_wordoptimized:
      MV      .S1     A6,A2               ;Count register
      ZERO    .S1     A9                  ;Sum register
||    ZERO    .S2     B9

loop: LDW     .D1     *A4++,A7            ;Load data from samples
                                          ;(here the input data is in 16 bit format)
||    LDW     .D2     *B4++,B7            ;Load data from coefficients

[A2]  SUB     .S1     A2,1,A2             ;Decrement counter
[A2]  B       .S1     loop
      NOP     2
      MPY     .M2     A7,B7,B8            ;B8 is the lower part product
||    MPYH    .M1     A7,B7,A8            ;A8 is the higher part product
      NOP
      ADD     .S1     A8,A9,A9            ;Accumulate result
||    ADD     .S2     B8,B9,B9            ;Accumulate result

      LDH     .D1     *A4++,A7            ;Load the final elements
||    LDH     .D2     *B4++,B7            ;Load the final elements
      NOP     4
      MPY     .M1     A7,B7,A8            ;Final multiply
      NOP
      ADD     .L1     A8,A9,A9            ;Final add

      ADD     .S1     A9,B9,A4            ;Move result to return register
      SHL             A4,1,A4             ;Eliminate sign extension bit

      B       .S2     B3                  ;Branch back to calling address
      NOP     5
```

Notice that since two loads are done consecutively, it takes half the amount of time to loop through the program. When calling this function, the value passed in A6 must be the truncated $N/2$, where N is the number of coefficients of the FIR filter. In our case, we have 11 coefficients requiring five iterations plus an additional multiply and accumulate. With this code, it is possible to bring down the number of cycles to 101. The timing cycles for the aforementioned optimizations are listed in Table 7-5.

Table 7-5: Timing cycles for different optimizations.

Optimization	Number of Cycles
Un-optimized assembly	313
Delay slot filled assembly	141
Word optimized assembly	101

L4.2.1 Handwritten Software-pipelined Assembly

To produce a software-pipelined version of the code, it is required to first write it in symbolic form without any latency or register assignment. The following code shows how to write the FIR program in a symbolic form:

```
    LDW        *p_sample++,sample   ;load sample word
    LDH        *p_coef++,coef       ;load coef half-word
    MPYHL      sample,coef,temp     ;temp = sample(high)*coef
    SHL        temp,1,temp          ;shift left to remove sign extended bit
    ADD        sum,temp,sum         ;sum += temp
[count] SUB       count,1          ;decrement counter
[count] B         loop             ;branch back to loop
```

To handwrite software-pipelined code, a dependency graph of the loop must be drawn and a scheduling table be created from it. The software-pipelined code is then derived from the scheduling table. To draw a dependency graph, we start by drawing nodes for the instructions and symbolic variable names. Then we draw lines or paths that show the flow of data between nodes. The paths are marked by the latencies of the instructions of their parent nodes.

After the basic dependency graph is drawn, functional units have to be assigned. Then, a line is drawn between the two sides of the CPU so that the workload is split as equally as possible. In the preceding FIR program, the loads should be done one on each side, so that they run in parallel. It is up to the programmer on which side to place the rest of the instructions to divide the workload equally between the A-side and B-side functional units. The completed dependency graph for the FIR program is shown in Figure 7-27.

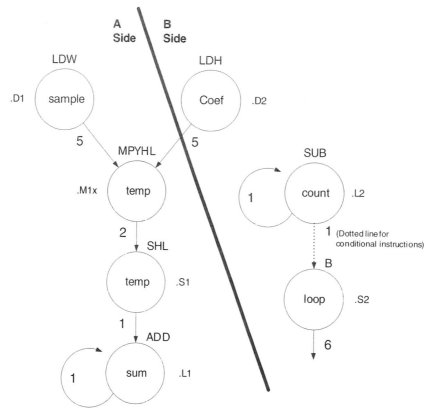

Figure 7-27: FIR dependency graph.[†]

The next step for handwriting a pipelined code is to set up a scheduling table. To do so, the longest path must be identified to determine how long the table should be. Counting the latencies of each side, one sees that the longest path located on the left side is 9. Thus, eight prolog columns are required in the table before entering the main loop. There need to be eight rows (one for each functional unit) and nine columns in the table. The scheduling is started by placing the parallel load instructions in slot 1. The instructions are repeated at every loop thereafter. The multiply instruction must appear five slots after the loads, so it is scheduled into slot 6. The shift must appear two slots after the multiply, and the add must appear after the shift instruction, placing it in slot 9, which is the loop kernel part of the code. The branch instruction is scheduled in slot 4 by reverse counting five cycles back from the loop kernel. The subtraction must occur before the branch, so it is scheduled in slot 3. The completed scheduling table appears in Table 7-6.

Table 7-6: FIR scheduling table.†

	PROLOG								LOOP
	1	2	3	4	5	6	7	8	9
.L1									ADD
.L2			SUB	SUB	SUB	SUB	SUB	SUB	SUB
.S1								SHL	SHL
.S2				B	B	B	B	B	B
.M1						MPYHL	MPYHL	MPYHL	MPYHL
.M2									
.D1	LDW	LDW	LDW	LDW	LDW	LDW	LDW	LDW	LDW
.D2	LDH	LDH	LDH	LDH	LDH	LDH	LDH	LDH	LDH

The software-pipelined code is handwritten directly from the scheduling table as eight parallel instructions before entering a loop that completes all the adds. The resulting code, with which the number of cycles is reduced to 72, is as follows:

```
        .global     _fir_pipelined

_fir_pipelined:
            ZERO    A10
            MV      A6,B2

            LDW     .D1     *A4++,A7
||          LDH     .D2     *B4++,B7

            LDW     .D1     *A4++,A7
||          LDH     .D2     *B4++,B7

            LDW     .D1     *A4++,A7
||          LDH     .D2     *B4++,B7
||  [B2]    SUB     .L2     B2,1,B2

            LDW     .D1     *A4++,A7
||          LDH     .D2     *B4++,B7
||  [B2]    SUB     .L2     B2,1,B2
||  [B2]    B       .S2     loop10

            LDW     .D1     *A4++,A7
||          LDH     .D2     *B4++,B7
||  [B2]    SUB     .L2     B2,1,B2
||  [B2]    B       .S2     loop10

            LDW     .D1     *A4++,A7
||          LDH     .D2     *B4++,B7
||  [B2]    SUB     .L2     B2,1,B2
||  [B2]    B       .S2     loop10
||          MPY     .M1     A7,B7,A8
```

```
           LDW    .D1    *A4++,A7
    ||     LDH    .D2    *B4++,B7
    || [B2] SUB   .L2    B2,1,B2
    || [B2] B     .S2    loop10
    ||     MPY    .M1    A7,B7,A8

           LDW    .D1    *A4++,A7
    ||     LDH    .D2    *B4++,B7
    || [B2] SUB   .L2    B2,1,B2
    || [B2] B     .S2    loop10
    ||     MPY    .M1    A7,B7,A8
    ||     SHL    .S1    A8,1,A9

loop10:    LDW    .D1    *A4++,A7
    ||     LDH    .D2    *B4++,B7
    || [B2] SUB   .L2    B2,1,B2
    || [B2] B     .S2    loop10
    ||     MPY    .M1    A7,B7,A8
    ||     SHL    .S1    A8,1,A9
    ||     ADD    .L1    A9,A10,A10

           MV     .L1    A10,A4
           B      .S2    B3
           NOP    5
```

L4.2.2 Assembly Optimizer Software-Pipelined Assembly

Since handwritten pipelined codes are time-consuming to write, linear assembly is usually used to generate pipelined codes. In linear assembly, latencies, functional units and register allocations do not need to be specified. Instead, symbolic variable names are used to write a sequential code with no delay slots (NOPs). The file extension for a linear assembly file is .sa. The assembly optimizer is automatically invoked if a file in a CCS project has a .sa extension. The assembly optimizer turns a linear assembly code into a pipelined code. Notice that the optimization level option in C also affects the optimization of linear assembly.

A code line in linear assembly consists of five fields: label, mnemonic, unit specifier, operand list, and comment. The general syntax of a linear assembly code line is:

```
[label[:]] [[register]] mnemonic [unit specifier] [operand list] [;comment]
```

Fields in square brackets are optional. A label must begin in column 1 and can include a colon. A mnemonic is an instruction such as MPY or an assembly optimizer directive such as .proc. Notice that a mnemonic will be interpreted as a label if it begins in column 1. At least one blank space should be placed in front of a mnemonic when there is no label. A mnemonic becomes a conditional instruction when it is preceded by a register or a symbolic variable within two square brackets. A unit specifier specifies the functional unit performing the mnemonic. Operands can be symbols, constants, or expressions and should be separated by commas. Comments begin with a semicolon. Comments beginning in column 1 can begin with either an asterisk or a semicolon.

In writing code in linear assembly, the assembly optimizer must be supplied with the right kind of optimization information. The first such piece of information is which parts should be optimized. The optimizer considers only code between the directives .proc and .endproc. Symbolic variable names are used to allow the optimizer to select which registers to use. This is done by using the .reg directive together with the names of the variables. Also, registers that contain input arguments, such as variables passed to a function, must be specified. The registers declared to contain input arguments cannot be modified and have to be declared as operands of the .proc statement. To connect the input register arguments to the symbolic variable names, the move instruction, mv, is used. Registers that contain output values upon exiting the procedure must be declared as arguments to the .endproc directive.

To write the FIR code in linear assembly, we start by creating the main loop and then add the load, multiply, and add instructions. Since two pointers to two arrays and an integer are passed to the function, we must declare registers A4, B4 and A6 as part of the .proc directive. Also, register A4 is used for returning values, so it needs to appear as part of the .endproc directive. The preserved register B3 is indicated as an argument in both of these directives. To connect the symbolic variable names to the input registers, the mv instruction is used. And finally, the optimizer is told that the loop is to be performed a minimum of 11 times by inserting the .trip directive. The final code is as follows:

```
        .global    _fir_la
        .sect  ".fir_la"

_fir_la:    .proc A4,B4,A6,B3
            .reg  p_m,m,p_n,n,prod,sum,cnt

            mv      A4, p_m           ;move argument in A4 to p_m
            mv      B4, p_n           ;move argument in B4 to p_n
            mv      A6, cnt           ;set up counter (third argument)
            zero    sum               ;sum=0

loop8:      .trip 11                  ;minimum 11 times through loop
            ldw     *p_m++, m         ;load m
            ldh     *p_n++, n         ;load n
            mpy     m,n,prod          ;prod = m * n
            shl     prod,1,prod       ;prod << 1
```

```
            add     prod,sum,sum        ;sum += prod
[cnt]       sub     cnt,1,cnt           ;decrement counter
[cnt]       b       loop8               ;branch back to loop8

            mv      sum,A4              ;move result into return register A4
            .endproc  A4,B3

            B       B3                  ;branch back to address stored in B3
            NOP     5
```

Using this code, we obtain a timing outcome of 72 cycles, which is the same as the timing obtained by the handwritten software-pipelined assembly.

To summarize the programming approach, start writing your code in C, and then use the optimizer to achieve a faster code. If the code is not as fast as expected, you may write it in assembly and incorporate the aforementioned simple optimization techniques. However, it is usually easier and more efficient to rewrite your code in linear assembly, since the assembly optimizer attempts to create pipelined code for you. Figure 7-28 illustrates the code development flow to get an optimum performance on the C6x. If, at the end, none of these approaches provide a satisfactory timing cycle, you are left no choice but to rewrite your code in hand-coded pipelined assembly. Appendix A (Quick Reference Guide) includes an optimization checklist for writing DSP application programs.

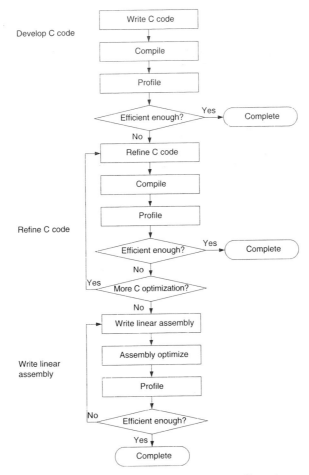

Figure 7-28: Code development flow.[†]

L4.3 Floating-Point Implementation

Implementing the FIR filter on the floating-point C67x takes relatively less effort. Since the hardware is capable of multiplying and adding floating point numbers, Q-format number manipulation is not required. However, in general, the floating-point code is slower, because floating-point operations have more latencies than their fixed-point counterparts. As is shown shortly, the FIR filter interrupt and function are modified for the floating-point execution. The code written in C is fairly simple. The coefficients are entered directly as float. The data buffer is declared as float, and a sample is initially read as an integer and then typecast to a float. Reversely, the outcome is typecasted into integer.

You can view your C source code interleaved with disassembled code in CCS. To view this mixed mode, after loading the program into the DSP, open the source file by double-clicking on it from the **Project View** panel. Select **View → Mixed Source/ASM** from the menu. Using this mixed mode, it can be verified that the compiler is actually using the MPYSP and ADDSP instructions to perform the floating-point multiply and add rather than calling a separate function to do them in software. The code is as follows:

```
float dotp1(const float a[], const float b[])
{
        int i;
        float sum = 0.0;
        for( i = 0 ; i < 11 ; i++ )
                sum += a[i] * b[i];
        return sum;
}

void interrupt serialPortRcvISR (void)
{
        int i, sample;
        float temp, sum;
        sample = MCBSP_read(hMcbsp);

        temp = (float)sample;
        for( i = 10 ; i >= 0 ; i-- )
                x[i] = x[i-1];
        x[0] = temp;
        sum = dotp1(coef, x);
        MCBSP_write(hMcbsp, (int)sum);
}
```

All the programs associated with this lab can be loaded from the accompanying CD.

L4.4 EVM Implementation

Since, on EVM, data are acquired from the codec through McBSP0, different than that of DSK, appropriate changes need to be made. As discussed in the preceding section, data samples from the left channel are stored in the higher 16-bit of integer variable. In order to maintain compatibility with the DSK code, the intrinsic_ MPYHL is used instead of _MPY for the multiplication of 32-bit samples with 16-bit coefficients.

```
interrupt void serialPortRcvISR(void)
{
        int i, temp, result;

        result = 0;

        temp = MCBSP_READ(0);
```

```
    // Update array samples
    for( i = N-1 ; i >= 0 ; i-- )
          samples[i+1] = samples[i];

    samples[0] = temp;

    // Filtering
    for( i = 0 ; i <= N ; i++ )
          result += ( _mpyhl(samples[i],b[i]) ) << 1;

    MCBSP_WRITE(0, result);
}
```

Alternatively, the sample in the DRR can be right shifted by 16 and stored in the lower 16-bit, followed by the MPY instruction, as shown in the code that follows:

```
interrupt void serialPortRcvISR(void)
{
      .
      .
      temp = MCBSP_READ(0);
      temp = temp >> 16;


      .
      .

    // Filtering
    for(i=0;i<=N;i++)
          result += ( _mpy(samples[i],b[i]) ) << 1;

    MCBSP_WRITE(0, result);
}
```

The output value stored in result contains Q-31 format data with the extra bit at the least significant bit. Data in the higher 16-bit are considered as Q-15 format and exported to the left channel, thus no alignment of data is necessary.

Bibliography

[1] The Mathworks, *MATLAB Reference Guide*, 1999.

[2] Texas Instruments, *TMS320C6000 Optimizing C Compiler User's Guide*, Literature ID# SPRU 187E, 1998.

CHAPTER 8

Circular Buffering

In many DSP algorithms, such as filtering, adaptive filtering, or spectral analysis, we need to shift data or update samples (i.e., we need to deal with a moving window). The direct method of shifting data is inefficient and uses many cycles. Circular buffering is an addressing mode by which a moving-window effect can be created without the overhead associated with data shifting. In a circular buffer, if a pointer pointing to the last element of the buffer is incremented, it is automatically wrapped around and pointed back to the first element of the buffer. This provides an easy mechanism to exclude the oldest sample while including the newest sample, creating a moving-window effect as illustrated in Figure 8-1.

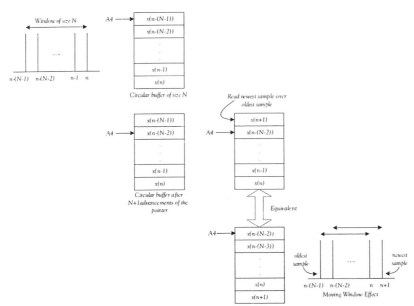

Figure 8-1: Moving-window effect.

Some DSPs have dedicated hardware for doing this type of addressing. On the C6x processor, the arithmetic logic unit has the circular addressing mode capability built into it. To use circular buffering, first the circular buffer sizes need to be written into the BK0 and BK1 block size fields of the Address Mode Register (AMR), as shown in Figure 8-2. The C6x allows two independent circular buffers of powers of 2 in size. Buffer size is specified as $2^{(N+1)}$ bytes, where N indicates the value written to the BK0 and BK1 block size fields.

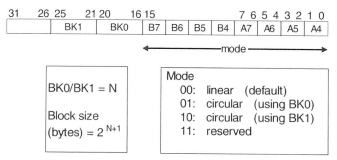

Figure 8-2: AMR (Address Mode Register).[†]

Then, the register to be used as the circular buffer pointer needs to be specified by setting appropriate bits of AMR to 1. For example, as shown in Figure 8-2, for using A4 as a circular buffer pointer, bit 0 or 1 is set to 1. Of the 32 registers on the C6x, 8 can be used as circular buffer pointers: A4 through A7 and B4 through B7. Note that linear addressing is the default mode of addressing for these registers.

Figure 8-3 shows the code to set up the AMR register for a circular buffer of size 8, together with a load example. To set up such a circular buffer in C, one must use so-called in-line assembly as follows:

```
asm ("MVK.S2      0001h,B2");
asm ("MVKLH.S2    0002h,B2");
asm ("MVC.S2      B2,AMR");
```

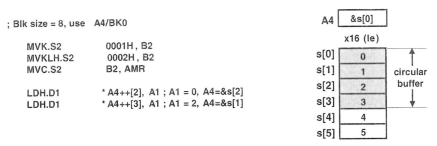

Figure 8-3: AMR setup example.[†]

When using circular buffers, care must be taken to align data on the buffer size boundary. In C, this can be achieved by using `pragma` directives. Pragma directives indicate what kinds of preprocessing are done by the compiler. The `DATA_ALIGN` `pragma` can be used to align *symbol* to a power of 2 alignment boundary *constant* (in bytes) as follows:

```
#pragma DATA_ALIGN (symbol,constant)
```

Lab 5: Adaptive Filtering

Adaptive filtering is used in many applications ranging from noise cancellation to system identification. In most cases, the coefficients of an FIR filter are modified according to an error signal in order to adapt to a desired signal. In this lab, a system identification example is implemented wherein an adaptive FIR filter is used to adapt to the output of a seventh-order IIR bandpass filter. The IIR filter is designed in MATLAB and implemented in C. The adaptive FIR is first implemented in C and later in assembly using circular buffering.

In system identification, the behavior of an unknown system is modeled by accessing its input and output. An adaptive FIR filter can be used to adapt to the output of the system based on the same input. The difference in the output of the system, $d[n]$, and the output of the adaptive filter, $y[n]$, constitutes the error term $e[n]$, which is used to update the coefficients of the FIR filter. Figure 8-4 illustrates this process.

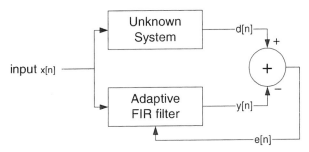

Figure 8-4: Adaptive filtering.

The error term calculated from the difference of the outputs of the two systems is used to update each coefficient of the FIR filter according to the formula (least mean square (LMS) algorithm [1]):

$$h_n[k] = h_{n-1}[k] + \delta * e[n] * x[n-k] \tag{8.1}$$

where the h's denote the unit sample response or FIR filter coefficients. The output $y[n]$ is required to approach $d[n]$. The term δ indicates step size. A small step size will ensure convergence, but results in a slow adaptation rate. A large step size, though faster, may lead to skipping over the solution.

L5.1 Design of IIR Filter

A seventh-order bandpass IIR filter is used to act as the unknown system. The adaptive FIR is designed to adapt to the response of the IIR system. Considering a sampling frequency of 8 kHz, let the IIR filter have a passband from $\pi/3$ to $2\pi/3$ (radians), with a stopband attenuation of 20dB. The design of the filter can be easily achieved with the MATLAB function 'yulewalk' [2]. The following MATLAB code may be used to obtain the coefficients of the filter:

```
Nc=7;
f=[0 0.32 0.33 0.66 0.67 1];
m=[0 0 1 1 0 0];
[B,A]=yulewalk(Nc,f,m);
freqz(B,A);

%Create A sample signal
Fs=8000;
Ts=1/Fs;
Ns=128;
t=[0:Ts:Ts*(Ns-1)];
f1=750;
f2=2000;%The one to keep
f3=3000;

x1=sin(2*pi*f1*t);
x2=sin(2*pi*f2*t);
x3=sin(2*pi*f3*t);

x=x1+x2+x3;
%Filter it
y=filter(B,A,x);
```

It can be verified that the filter is working by deploying a simple composite signal. Using the MATLAB function 'filter', verify the design by observing that the frequency components of the composite signal falling in the stopband are removed. (See Figure 8-5 and Table 8-1.)

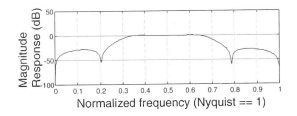

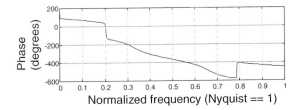

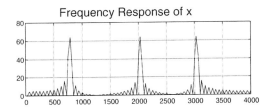

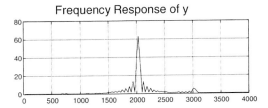

Figure 8-5: IIR filter response.

Table 8-1: IIR filter coefficients.

A's	B's
1.0000	0.1191
0.0179	0.0123
0.9409	−0.1813
0.0104	−0.0251
0.6601	0.1815
0.0342	0.0307
0.1129	−0.1194
0.0058	−0.0178

Note: Do not confuse A&B coefficients with the CPU A&B registers!

L5.2 IIR Filter Implementation

The implementation of the IIR filter is first done in C, using the following difference equation

$$y[n] = -\sum_{k=1}^{N} a_k * y[n-k] + \sum_{k=0}^{N} b_k * x[n-k],$$ (8.2)

where a_k's and b_k's denote the IIR filter coefficients. Two arrays are required, one for the input samples $x[n]$ and the other for the output samples $y[n]$. Given that the filter is of the order 7, an input array of size 8 and an output array of size 7 are considered. The arrays are used to simulate a circular buffer, since in C this property of the CPU cannot be accessed. As a new sample comes in, all elements in the input array are shifted down by one. In this manner, the last element is lost and the last eight samples are always kept. The input array is used to calculate the resulting output, and then the output is used to modify the output array. A simple implementation of this scheme is shown in the following code, which is a modification of the sampling program in Lab 2:

```
interrupt void serialPortRcvISR (void)
{
        int temp,n,ASUM,BSUM;
        short input,IIR_OUT;

        temp = MCBSP_read(hMcbsp);
        input = temp >> S;                          //Scaling factor

        for(n=7;n>0;n--)                            //Input buffer
                IIRwindow[n] = IIRwindow[n-1];

        IIRwindow[0] = input;

        BSUM = 0;

        for(n=0;n<=7;n++)
        {                                           //Multiplication of Q-15 with Q-15
         BSUM += ((BS[n]*IIRwindow[n]) << 1);       //Results in Q-30.Shift by one to
        }                                           //Eliminate Sign Extension bit

        ASUM = 0;
        for(n=0;n<=6;n++)
        {
                ASUM += ((AS[n] * y_prev[n]) << 1);
        }

        IIR_OUT = (BSUM - ASUM) >> 16;

        for(n=6;n>0;n--)                            //Output buffer
                y_prev[n] = y_prev[n-1];

        y_prev[0] = IIR_OUT;

        MCBSP_write(hMcbsp, IIR_OUT << S);          // Scaling factor S
}
```

By running this program while connecting a function generator and an oscilloscope to the line-in and line-out of the audio daughter card, the functionality of the IIR filter can be verified. Whenever the DRR receives a new incoming sample from the function generator, the ISR is invoked. Then, the new sample is right shifted by the scaling factor S. This factor is included for the correction of any possible overflow. In this lab, there is no need for shifting. Once the new sample is scaled, the last eight samples are kept by discarding the oldest sample and adding the new sample to the input buffer IIRwindow. This operation is done by shifting the data in the input array. Note that this array is global and is initialized to zero in the main function.

Now that the last eight samples are ready to be used, it is time to compute BSUM (*b* coefficient terms) and ASUM (*a* coefficient terms). Attention needs to be paid to the datatype of BSUM, ASUM, the coefficient arrays, and the input array. The datatype of the coefficient arrays is short, so the coefficients are converted to Q-15 format by multiplying them by 0x7FFF in the main function. The datatype of the input array IIRwindow is also short. However, the datatype of ASUM or BSUM is int (32 bits). Therefore, ASUM and BSUM need to be left shifted by 1 to eliminate the extended sign bit, since the multiplication of Q-15 by Q-15 results in Q-30 representation. In order to obtain the IIR output IIR_OUT, the partial output ASUM is subtracted from BSUM. Note that the difference (BSUM – ASUM) is right shifted by 16 to convert it to a short datatype. The IIR output is then used to compute ASUM in the next iteration. Finally, the IIR output is scaled back and sent to the data transmit register (DXR).

L5.3 Adaptive FIR Filter

By replacing the following piece of code with MCBSP_write(hMcbsp,IIR_OUT<<S) in the previous IIR function, we can make a FIR filter to adapt to the output of the IIR filter:

```
//Simulate Circular buffer for FIR
for(n=31;n>0;n--)
      FIRwindow[n] = FIRwindow[n-1];

FIRwindow[0] = input;

//Perform Filtering with current coefficients
temp = 0;
for(n=0;n<32;n++)
{
      temp += ((h[n]*FIRwindow[n]) << 1);
}
```

```
y = temp >> 16;

//Calculate Error Term

e = IIR_OUT - y;

//Update Coefficients

stemp = (DELTA*e)>>15;

for(n=0;n<32;n++)
{
        stemp2 = (stemp*FIRwindow[n])>>15;
        h[n] = h[n] + stemp2;
}
MCBSP_write(hMcbsp,y<<S);
```

In this program, a 32-coefficient FIR filter is used to adapt to the output of the IIR filter. To do this in C, an additional buffer of length 32 is needed: one for the input buffer FIRwindow and the other for the coefficients h of the FIR filter. Initially all the data in both arrays are zero. The order of processing is as follows: First, the last 32 samples are shifted. The shift discards the oldest sample and adds the newly read sample into the input buffer. Next, the FIR filtering is done by performing a dot-product between the coefficients h and the input buffer. The dot-product is converted to Q-15 format by left shifting it by 1. Then, the error term between the IIR and the FIR filter output is computed. The coefficients of the FIR filter are updated to match the IIR and FIR filter outputs. Finally, the FIR filter output is sent to the DXR. By using a function generator and an oscilloscope, the adaptation process can be observed by scanning through different frequencies.

It is worth mentioning a point about the step size δ. In a floating-point processor, δ is usually chosen to be in the range of e^{-7}. However, the precision on the fixed-point C6x does not allow for such a small number. We can at most use 0x0001, which is $1/(2^{15}) \approx 0.0000305$. When a multiplication is done with this number, any positive number will be defaulted to 0 and any negative number to -1. This is due to the nature of multiplication of Q-15 format numbers, where the product is right shifted by 15. However, the contribution of negative numbers to the coefficients is sufficient for the LMS algorithm to converge. Using a larger δ for this adaptive filtering example results in faster adaptation, but convergence is not guaranteed. Satisfactory results can be observed with δ in the range of 0x0100 to 0x0001.

The reason for implementing the LMS algorithm in assembly is to make use of the circular buffering capability of the C6x. Of the 32 registers on the C6x, 8 can perform circular addressing. These registers are A4 through A7 and B4 through B7. Since linear addressing is the default mode of addressing, each of these registers must be specified as circular using the AMR register. The lower 16 bits of the AMR register are used to select the mode for each of the 8 registers. The upper 10 bits (6 are reserved) are used to set the length of the circular buffer. Buffer size is determined by $2^{(N+1)}$ bytes, where N is the value appearing in the block size fields of the AMR register.

Since we are using both C and assembly, we have to initialize the circular buffer when we enter the assembly part of the program. During the execution of the assembly code, the register used in the circular mode allows a certain location in memory to always contain the newest sample. As the assembly code is completed and returns to the calling C program, the location of the pointer to the buffer must be saved and the AMR register must be returned to the linear mode, since leaving it in the circular mode disrupts the flow of the program.

To do such a task, a section of memory not used by the compiler must be set aside for the buffer, and the coefficients. A simple way to do this is to reserve 64 bytes for the coefficients, 64 bytes for the buffer, and 4 bytes for the pointer. Since the data and coefficients are short formatted here, 64 bytes are used to provide 32 locations. The following memory representation is employed for this purpose:

0x00000200	64 Bytes, Circular Buffer
0x00000240	64 Bytes, Coefficients
0x00000280	4 Bytes, Pointer

The command file must also be modified. A simple assembly file is needed to initialize the memory locations with zeros. The following command file defines a new memory section called MMEM in the internal memory and uses it for the code section .mydata:

```
MEMORY
{
        vecs:       o = 00000000h    l = 00000200h
        MMEM:       o = 00000200h    l = 00000100h
        IRAM:       o = 00000300h    l = 0000FD00h
        CE0:        o = 80000000h    l = 01000000h
```

```
SECTIONS
{
        "vectors"    >      vecs
        .cinit       >      IRAM
        .text        >      IRAM
        .stack       >      IRAM
        .bss         >      IRAM
        .const       >      IRAM
        .data        >      IRAM
        .far         >      IRAM
        .switch      >      IRAM
        .tables      >      IRAM
        .cio         >      IRAM
        .sysmem      >      CE0
        .mydata      >      MMEM
}
```

The file *initmem.asm* appearing next is used to initialize the memory locations with zeros and set the pointer to the first free location, which is $0x00000200$:

```
; initmem.asm

    .sect ".mydata"
    .short 0,0,0,0,0,0,0,0,0,0,0,0,0,0,0,0,0,0,0,0,0,0,0,0,0,0,0,0,0,0,0,0
    .short 0,0,0,0,0,0,0,0,0,0,0,0,0,0,0,0,0,0,0,0,0,0,0,0,0,0,0,0,0,0,0,0
    .field 0x00000200, 32
```

With the command file and the brief assembly code just shown, it is ensured that 132 bytes of space starting at $0x00000200$, will not be used for anything other than the adaptive filter. Now, as mentioned before, the circular buffer must be initialized upon entering the assembly part. To do this, it is necessary to modify the AMR register. Since a buffer of length 32 is needed, one must have 5 in the block fields (block size $= 2^{(5+1)} = 64$). With register A5 as the circular buffer pointer, the value to set the AMR register becomes $0x00050004$. Entering the assembly function, the last pointer location is read from $0x00000280$. The last free location of the buffer is saved to the same location upon exit. The following code shows how this is achieved:

```
     ;Initialize the Circular buffer for the FIR filter
     MVK    .S2    0x0004,B10         ;A5 is selected as circular
     MVKLH .S2     0x0005,B10         ;2^(5+1)=64
     MVC    .S2    B10,AMR

     ;Load the pointer to A0
     ;Assume that the current location of the circular buffer is pointed to

     MVK    .S1    0x0280,A0
     MVKLH .S1     0x0000,A0          ;A0=0x00000280 (has last pointer)
```

```
     LDW    .D1    *A0,A5              ;A5 now points to the first free
                                       ;Location of the circular buffer

     NOP    4

     ;Load the current sample to the Circular buffer
     STH    .D1    A4,*A5             ;A4 has sample passed from calling


     //FIR FILTERING HERE

     ;Now save the Last location of A0 to memory

     MVK    .S1    0x0280,A0
     MVKLH  .S1    0x0000,A0          ;This address has the pointer to x


     LDH    .D1 *A5++,A13             ;Dummy Load
     STW    .D1 A5,*A0                ;Saved last
     ;Restore Linear Addressing
     MVK    .S2    0x0000,B10
     MVKLH  .S2    0x0004,B10
     MVC    .S2    B10,AMR


     ;return the result y

     MV     .S1    A9,A4
     B      .S2    B3
     NOP    5
```

Upon entering the assembly function, the AMR register is loaded with 0x00050004 for the desired circular buffer operation. Then, the memory location (or address) to which register A5 was last pointing is loaded into A5. Hence, A5 points to the first free location of the circular buffer, and the content of register A4 (the newest sample passed from the C program) is stored in this location. After adaptive filtering, the address pointed to by A5 is stored at the location 0x00000280. Note that here a dummy load is performed to increment the pointer so that it points to the last element (the next free location). This is needed because only a load or store operation increments the pointer in a circular fashion.

The following adaptive FIR filter assembly code resides in the section of the foregoing code labeled 'FIR FILTERING HERE':

```
     ;Do the filtering
     MVK    .S2    0x0240,B1
     MVKLH  .S2    0x0000,B1          ;This is the address of h[n]

     MVK    .S2    32,B2              ;Set up a counter
     ZERO   .S1    A9                 ;Accumulator

loop:
     LDH    .D2    *B1++,B7           ;load h_x
     LDH    .D1    *A5--,A7
```

```
         NOP     4
         MPY     .M1x   A7,B7,A7              ;A7 is Q-30
         NOP
         SHL     .S1    A7,1,A7
         ADD     .S1    A7,A9,A9
[B2]     SUB     .S2    B2,1,B2              ;Decrement Counter
[B2]     B       .S2    loop
         NOP     5

         SHR     .S1    A9,16,A9             ;Make Short, Eliminate Sign extension bit
                                             ;A9 is now short Y
         ;Calculate Error Term
         MV      .S1    B4,A13
         SUB     .S1    A13,A9,A8            ;A13=d(IIR_OUT),A9=y,A8=e

         ;Update Coefficients
         MVK     .S1    0x0001,A10           ;A10=DELTA
         MPY     .M1    A8,A10,A10           ;A10=DELTA*e this is actually ineffective
         NOP
         SHR     .S1    A10,15,A10           ;A10=DELTA*e is now short Q-15

         MVK     .S2    32,B2                ;Loop Counter
         MVK     .S2    0x0240,B1
         MVKLH   .S2    0x0000,B1            ;This is the address of h[n]

loop2:
         LDH     .D1    *A5--,A8             ;Load x[n-k]
         LDH     .D2    *B1,A12              ;Load h[n]
         NOP     4
         MPY     .M1    A10,A8,A8            ;A10 = DELTA*e*x in Q-31
         NOP
         SHR     .S1    A8,15,A8             ;A10 is now Q-15
         ADD     .S1    A8,A12,A8            ;Updated h
         STH     .D2    A8,*B1++             ;Update the coefficient
[B2]     SUB     .S2    B2,1,B2              ;Decrement Counter
[B2]     B       .S2    loop2
         NOP     5
```

In this code, the adaptive filtering process is done via two separate loops. The first loop calculates a dot-product between the coefficients and the samples. The error term is then calculated and used in the second loop for updating the coefficients, based on the input samples. Notice that a circular buffer is not used for the coefficients, since they do not change in a time-windowed manner, as do the input samples.

We now have two versions of the adaptive filter. One is written entirely in C, and the other is a mix of C and assembly. When the entire C program runs in the external memory, the output does not adapt to the IIR filter output. Only when the entire C program runs in the internal memory does the output adapt to the IIR filter output. Also, when the assembly part of the mixed C/assembly program runs in the external memory, the output adapts to the IIR filter output. Of course, the assembly part of the program can be configured to run in the internal memory space, where the number of cycles is noticeably reduced. The main reason for running the C code from the internal memory space is that running it from the external memory is too slow and samples get missed. Table 8-2 gives a summary of the timing cycles for different memory options of the adaptive filtering program. All the programs associated with this lab are placed on the accompanying CD-ROM.

Table 8-2: Timing cycles for different memory options.

Type of build	Number of cycles
C program in external memory	22480
C program in internal memory	3083
Non-optimized assembly in external memory	10009
Non-optimized assembly in internal memory	1200

Bibliography

[1] S. Haykin, *Adaptive Filter Theory*, Prentice-Hall, 1996.

[2] The Mathworks, *MATLAB Reference Guide*, 1999.

CHAPTER 9

Frame Processing

When it comes to processing frames of data (for example, in doing FFT and block convolution), triple buffering is an efficient data frame handling mechanism. While samples of the current frame are being collected by the CPU in an `input` array via an ISR, samples of the previous frame in an `intermediate` array can get processed during the time left between samples. At the same time, the DMA can be used to send out samples of a previously processed frame available in an `output` array. In this manner, the CPU is used to set up the `input` array and process the `intermediate` array while the DMA is used to move processed data from the `output` array. At the end of each frame or the start of a new frame, the roles of these arrays are interchanged. The `input` array is reassigned as the `intermediate` array to be processed, the processed `intermediate` array is reassigned as the `output` array to be sent out, and the `output` array is reassigned as the `input` array to collect incoming samples for the current frame. This process is illustrated in Figure 9-1.

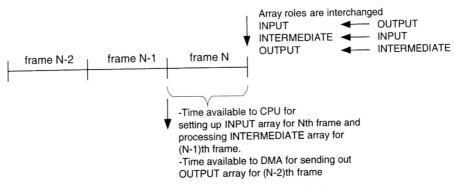

Figure 9-1: Triple buffering technique.

9.1 Direct Memory Access

Many DSP chips are equipped with a Direct Memory Access resource acting as a co-processor to move data from one part of memory into another without interfering with the CPU operation. As a result, the chip throughput is increased since, in this manner, the CPU can process and the DMA can move data without interfering with each other.

Depending upon the DSP platform used, there are two different DMAs called DMA and EDMA (enhanced DMA). The differences between them and their configurations are stated in the following subsections.

9.1.1 DMA

The C6x01 DSP provides four DMA channels. Each DMA channel has its own memory-mapped control registers which can be set up to move data from one place to another place in memory. Figure 9-2 shows the DMA control registers consisting of the Primary Control, Secondary Control, Source Address, Destination Address, and Transfer Counter registers. These registers contain the information regarding source and destination locations in memory, number of transfers, and format of transfers. As shown in Figure 9-2, in addition to the DMA control registers, there are several global registers shared by all DMA channels.

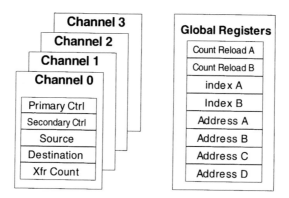

Figure 9-2: DMA channels control registers.[†]

An example is presented here to show how some of the fields of the DMA registers are set for block or frame processing. More details on all the fields are available in the *TI TMS320C6x Peripherals Reference Guide* [1]. It is possible to transfer a block of data consisting of a number of frames, which in turn consist of a number of elements. Elements here mean the smallest piece of data. The example shown in Figure 9-3 illustrates the DMA register setup for transferring a block of data (this can be viewed as image data) consisting of four frames (rows), while each frame consists of four elements (16-bit pixels). Figure 9-3 also provides the options for some of the fields of the Primary Control Register, Transfer Counter Register, and Global Index Register. Source/destination address fields can be incremented or decremented by an element size or by an index as specified in the global register. The element size field is used to indicate the datatype. The Transfer Counter Register contains the number of frames and elements. In this example, the element size field of the Primary Control Register is set to 01 (halfwords), the source address field to 11 (index option for accessing next value), and the destination to 01 (increment option for writing consecutively). As specified in the Transfer Counter Register, the data transfer in this example consists of four frames, and each frame consists of four elements. The global register A (chosen by the index field) contains the element as well as the frame index. In this example, the element index is set to 2 to increase addresses by 2 bytes or element size, and the frame index to 6 bytes to get to the next frame, since at the end of a frame the pointer points to the beginning of the last element of that frame.

16-bit Pixes

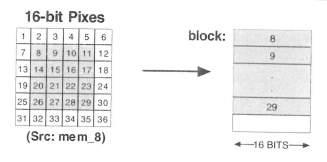

(Src: mem_8)

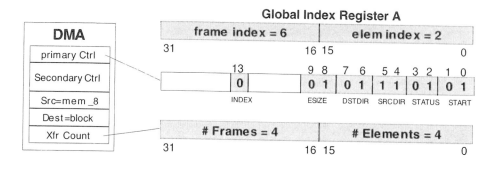

ESIZE		SRC/DST DIR		STATUS		START	
00	32-bits	00	no modifications	00	Stopped	00	Stop
01	16-bits	01	inc by element size	01	Running (without auto-init)	01	Start without auto-init
10	8-bits	10	dec by element size	10	Paused	10	Pause
11	reserved	11	index	11	Running (with auto- init)	11	Start with auto-init

Figure 9-3: DMA data transfer example.†

9.1.2 EDMA

C6711/C6713/C6416 DSKs possess EDMA. The number of EDMA channels are 16 on C671X, and 64 on C64x. As compared with DMA, EDMA provides programmable priority, and the ability to link data transfers. By using the EDMA controller, data can be transferred from/to internal memory (L2 SRAM), or peripherals, to/from external memory spaces efficiently without interfering with the CPU operation. Typically, block data transfers and transfer requests from peripherals are performed via EDMA. Figure 9-4 illustrates the EDMA registers used for its configuration.

31	16 15	0
Options (OPT)		Word 0
Source Address (SRC)		Word 1
Array/frame count (FRMCNT)	Element count (ELECNT)	Word 2
Destination address (DST)		Word 3
Array/frame index (FRMIDX)	Element index (ELEIDX)	Word 4
Element count reload (ELERLD)	Link address (LINK)	Word 5

Figure 9-4: EDMA registers.[†]

Also, EDMA has a feature called quick DMA (QDMA), which provides a fast and efficient way to transfer data for applications such as data requests in tight loop algorithms. QDMA allows single, independent transfers for quick data movement, rather than periodic or repetitive transfers, as done by EDMA channels. Figure 9-5 illustrates an example of 2-D to 1-D data transfer. A 16×12 subframe of a 640×480 video data (16-bit pixels), stored in the external memory, is transferred to the internal L2 RAM by using QDMA. Similar to the DMA example, the array index denotes the space between the subframe arrays. Thus, the array index is $2 * (640 - 16) = 1248$ (4E0h). In the QDMA Options register, the priority level is set to be low, element size to 16-bit half-word. The source is specified as two-dimensional with address increment, and the destination is specified as one-dimensional with address increment. The transfer complete interrupt (TCINT) field is disabled and the channel is synchronized by setting the frame synchronization field (FS) to 1. For a more detailed description of EDMA and QDMA, refer to *Applications Using the TMS320C6000 Enhanced DMA* [2].

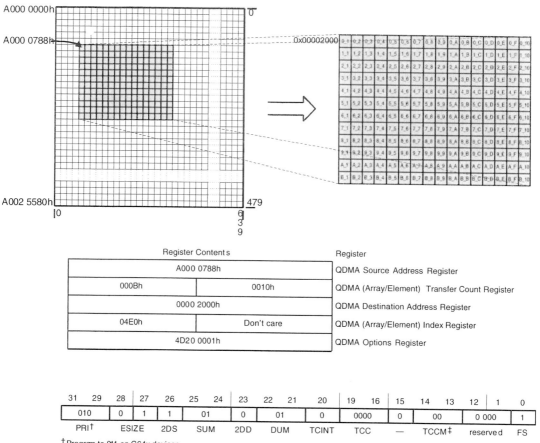

Register Contents		Register
A000 0788h		QDMA Source Address Register
000Bh	0010h	QDMA (Array/Element) Transfer Count Register
0000 2000h		QDMA Destination Address Register
04E0h	Don't care	QDMA (Array/Element) Index Register
4D20 0001h		QDMA Options Register

31 29	28	27	26	25 24	23	22 21	20	19 16	15	14 13	12 1	0
010	0	1	1	01	0	01	0	0000	0	00	0 000	1
PRI†	ESIZE	2DS		SUM	2DD	DUM	TCINT	TCC	—	TCCM‡	reserved	FS

† Program to 011 on C64x devices.
‡ Available only on C64x devices.

Figure 9-5: Subframe transfer example with QDMA.†

9.2 DSP-Host Communication

9.2.1 Host-Port Interface (HPI) on C6711 DSK

The HPI on C6711 DSK is a 16-bit-wide parallel port through which a PC host can directly access the DSP memory and peripherals. No specific configuration is required for the HPI communication on the DSP side since the host program acts as master.

A program on the PC host is written here in Microsoft® Visual C++® using the library *dsk6x11hpi.lib*, which is a part of CCS. In this program, a sample configuration file, *ti_ppdsk.cfg*, is accessed to open the device handle. After opening the DSK and the HPI with the `dsk6x_open()` and `dsk6x_hpi_open()` APIs, respectively, the memory of the DSP is read by the `dsk6x_hpi_read()` API based on the length and starting address parameters. This program appears below.

```
char *pfname = "c:\\ti\\c6000\\dsk6x11\\doc\\ti_ppdsk.cfg";

dsk6x_open(pfname, &handle);
dsk6x_hpi_open(handle);
dsk6x_hpi_read(handle, temp, &len, StrtAddr);
dsk6x_hpi_close(handle);
dsk6x_close(handle);
```

In order to build this program, the library *dsk6x11hpi.lib* should be included in the project and the Win32 DLL *dsk6x11hpi.dll* should be placed in the same folder where the program is located.

Bearing in mind that CCS already uses the HPI to load and run the program, the access should be returned before the host program runs on the PC. Thus, CCS should be closed before the host program is executed. The DSP target program is then loaded and run by the host program instead of CCS.

9.2.2 HPI on C6701 EVM

On the C6701 EVM, there are two 32-bit FIFO (first in, first out) registers, called FIFO-read and FIFO-write, in the PCI controller of the EVM board through which the PC host and the C6x can communicate. The FIFO-read is used for data transfer from the host to the DSP and the FIFO-write for data transfer from the DSP to the host. The initialization of the FIFO is done through the memory mapped FIFO status register. The communication can be simply achieved by using the supplied `Write-FIFO_DMA()` function as part of Lab6.

9.2.3 Real-Time Data Exchange (RTDX)

Since the HPI on the C6713/C6416 DSKs requires an external interface through the PCI/HPI port, RTDX is used here as an alternative way to communicate between the host and the DSP. On these platforms, CCS communicates with the DSK through an embedded JTAG emulator with a USB host interface.

RTDX details are discussed in Chapter 10. To use RTDX, first a channel needs to be created by calling the function RTDX_CreateOutputChannel(). Then, data can get transferred to the host in real-time by calling the RTDX_write() API, as shown below.

```
RTDX_CreateOutputChannel(ochan);
RTDX_enableOutput(&ochan);

// Send the data to the host
if ( !RTDX_write( &ochan, data, sizeof(data) ) ) {
    fprintf(stderr, "\nError: RTDX_write() failed!\n");
    abort();
}

// Wait for Target-to-Host transfer to complete
while ( RTDX_writing != NULL ) {
        #if RTDX_POLLING_IMPLEMENTATION
          RTDX_Poll();
        #endif
}

RTDX_disableOutput(&ochan);
```

Bibliography

[1] Texas Instruments, *TMS320C6201/6701 Peripherals Reference Guide*, Literature ID# SPRU 190B, 1998.

[2] Texas Instruments, *Applications Using the TMS320C6000 Enhanced DMA*, Literature ID# SPRA636A, 2001.

Lab 6: Fast Fourier Transform

Operations such as DFT or FFT require that a frame of data be present at the time of processing. Unlike filtering, where operations are done on every incoming sample, in frame processing N samples are captured first and then operations are performed on all N samples.

To perform frame processing, a proper method of gathering data and of processing and sending data out is required. The processing of a frame of data is not usually completed within the sampling time interval, rather it is spread over the duration of a frame before the next frame of data is gathered. Hence, incoming samples must be stored into a separate buffer other than the one being processed. Also, another buffer is needed to send out a previously processed frame of data. As explained earlier, this can be achieved by triple buffering involving three buffers: input, intermediate, and output.

To do triple buffering on the C6x, the sampling shell program in Lab 2 is modified to incorporate an endless loop revolving around the rotation of three buffers. The buffers rotate every time the input buffer is full so that a new frame of N sampled data is passed to the intermediate buffer for processing and a previously processed frame is passed to the output buffer for transmission. The following modifications of the shell program achieve this:

```
short *output;              /* POINTER TO DATA ARRAY FOR OUTPUT        */
short *input;               /* POINTER TO DATA ARRAY FOR INPUT         */
short *intermediate;        /* POINTER TO DATA ARRAY FOR DMA ACCESS    */
static short index=0;

main()
{
    CSL_init();                 // Initialize the library

    hMcbsp = MCBSP_open(MCBSP_DEV1,MCBSP_OPEN_RESET);
    MCBSP_config(hMcbsp,&MyConfig);

    hTimer = TIMER_open(TIMER_DEV0, TIMER_OPEN_RESET);
    TIMER_config(hTimer, &timerCfg);

    init_arrays();

    IRQ_globalDisable();
    IRQ_nmiEnable();
```

```
        IRQ_map(IRQ_EVT_RINT1,15);
        IRQ_enable(IRQ_EVT_RINT1);
        IRQ_globalEnable();

        /* Main Loop, wait for Interrupt */

        for( ; ; )
        {
                wait_buffer();        /* WAIT FOR A NEW BUFFER OF DATA */
        }
}
```

The preceding code shows how an endless loop is added to the shell program. Here, most of the initializations for the codec and McBSP have not been shown to make the code easier to follow. Once the serial port is initialized, the three arrays are allocated in memory and initialized to zero. The program then goes into an endless loop where the function wait_buffer(), shown next, is executed endlessly:

```
void wait_buffer(void)
{
        short *p;
        /* WAIT FOR ARRAY INDEX TO BE RESET TO ZERO BY ISR  */
        while(index);

        /* ROTATE DATA ARRAYS */
        p = input;
        input = output;
        output = intermediate;

        //Function call here...

        intermediate = p;
        HostTargetComm();
        while(!index);
}
```

This function checks on the global variable index to do the rotation of the arrays and to start processing. When the input array becomes full (indicated by index), the arrays are rotated and the intermediate array gets set for processing. The comment //Function call here... indicates where the processing function such as FFT should be placed.

The ISR is also modified as shown in the next code block. In the ISR, EDMA is used to transfer a sample from DRR to DXR without using the MCBSP_write() API. Note that index is incremented within the ISR.

```
EDMA_Handle hEdma;

EDMA_Config myConfig = {
        0x28000000, // opt
```

```
        0x01900000, // src: DRR 1
        0x00000001, // cnt
        0x01900004, // dst: DXR 1
        0x00000000, // idx
        0x00000000  // rld
};

interrupt void serialPortRcvISR (void)
{
        int temp;
        temp =  MCBSP_read(hMcbsp);
        input[index] = temp;

        myConfig.src = EDMA_SRC_RMK(&temp); // Update EDMA source

        hEdma = EDMA_open(EDMA_CHA_XEVT1, EDMA_OPEN_RESET);
        EDMA_config(hEdma,&myConfig);
        EDMA_enableChannel(hEdma);
        EDMA_setChannel(hEdma);
        EDMA_disableChannel(hEdma);
        EDMA_close(hEdma);


        if (++index == BUFFLENGTH)
            index = 0;
}
```

To configure EDMA for reading from DRR and writing back to DXR, the source and destination address should refer to DRR1 and DXR1, respectively. In the Options Parameter register, the priority level is set to high, element size to 32-bit, with one-dimensional source and destination. No incrementing of address for source or destination is necessary since the source and destination are fixed-address memory mapped registers. Similar to the other CSL peripherals, EDMA is opened and configured with the EDMA_open() and EDMA_config() APIs. The EDMA_setChannel() API triggers the EDMA channel. After transmitting data, the opened channel should be closed with EDMA_close().

For FFT processing purposes, the data read from DRR are left-shifted by 16 bits to get rid of the data portion corresponding to the right channel. To store in Q-15 format, it is shifted back right by 16 bits. The input samples are then placed into the input array to build a frame of length BUFFLENGTH. The variable index is incremented every time a new input sample is read. This variable is reset to zero when the input buffer becomes full, that is, index reaches BUFFLENGTH. This reset causes the program to go out of the while(index) loop in the function wait_buffer(). Then, the rest of the program in wait_buffer() gets executed.

Now, let us go back to wait_buffer(). After index is reset to zero, the input buffer is reassigned to a pointer named p. This reassignment is necessary for the

//Function call here... part to avoid any conflict with the ISR. Note that the ISR uses the input buffer to receive new samples. If say the FFT function processes the data in the input buffer, wrong results may be produced, because the ISR may change the content of the input buffer anytime. This malfunction may occur because the ISR runs on a higher priority basis, while the FFT function runs on a lower priority basis. Following the p=input statement, the output buffer is reassigned to the input buffer. This reassignment allows the ISR to use the output buffer to receive and store new incoming samples. Notice that the data in the output buffer is sent out by the DMA. The next reassignment output = intermediate is necessary in order to send the processed data in the intermediate buffer to the DXR.

After the data is processed in //Function call here..., the pointer p is reassigned to the intermediate buffer for it to point to the processed samples. The data in the intermediate buffer is sent out by the function Host-TargetComm() as part of the wait_buffer() function. The while loop while(!index) at the end of wait_buffer() ensures that a frame is processed only once. In the absence of a new sample in the DRR, index remains zero and the program waits at while(!index) because !index is TRUE. When a new sample arrives in the DRR, index is incremented and the program gets out of wait_buffer() and falls into the loop in main(). There it waits for a new frame.

For communication with a PC host, two different methods are mentioned here depending on the target platform (DSK or EVM) used. When using EVM, the communication is done through the memory mapped I/O address, PCI FIFO, corresponding to the PC PCI channel. There are two 32-bit FIFO (first in first out) registers, called FIFO-read and FIFO-write, in the PCI controller of the EVM board through which the PC host and the C6x may communicate. The FIFO-read is used for data transfer from the PC host to the DSP and the FIFO-write for data transfer from the DSP to the PC host. The initialization of the FIFO is done through the memory mapped FIFO status register. The communication is achieved by using the HostTargetComm() function. This function utilizes the C6x's DMA capability to send the intermediate array to the host through the FIFO registers. The following provides the code for doing so:

```
void HostTargetComm(void)
```

```
{
    dma_reset();
    dma_init(2,                    //Channel
             0x0A000110u,          //Primary Control Register    (Peripherals pp4-9)
             0x0000000Au,          //Secondary Control Register
(unsigned int) intermediate,       //Source Address
             0x01710000u,          //Destination Address
             0x00010080u);         //Transfer Counter Register
    DMA_START(DMA_CH2);
}
```

The two DMA API functions dma_reset() and dma_init() are used to initialize the DMA for data transfer between the intermediate array and the FIFO. A single frame of samples of length 128 is transferred as indicated by the content of the Transfer Counter Register (TCR). The dma_reset() API resets all the DMA registers to their power-on reset states. The dma_init() API initializes a selected DMA channel by assigning appropriate values to its control registers. Here, the first argument of the dma_init() API is used to select the DMA channel 2. The second argument sets the Primary Control Register to 0x0A000110u, as shown in Figure 9-6. This value is specified based upon the following: EMOD bit is set to 1 to pause the DMA channel during an emulation halt. TCINT bit is set to 1 to enable the transfer controller interrupt. The element size is 16 bits, so ESIZE bits are set to 01. SRC DIR bits are set to 01 to increment the source address by element size in bytes. The third argument of the dma_init() API sets the Secondary Control Register. SX IE and FRAME IE bits of this register are set to 1 to enable the DMA channel interrupt. The fourth argument of the dma_init() API assigns the intermediate array as the source address. The fifth argument is set to 0x01710000u, which is the address of the EVM PCI interface, through which the host program reads the output. The last argument sets the Transfer Counter Register to 0x00010080u. The upper 16 bits specify the number of frames. Here, these bits are set to 0x0001u in order to send out one frame at a time. The lower 16 bits are set to 0x0080u for having 128 elements in a frame. Finally, the function Host-TargetComm() calls the macro DMA_START() to activate the DMA channel 2. This macro sets the START field of the Primary Control Register to 01.

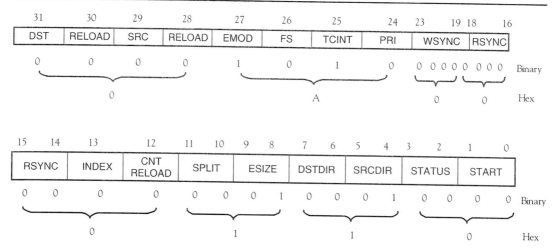

Figure 9-6: Primary Control Register.

When using C6711 DSK, the communication is done through the parallel port. In order to implement a function equivalent to `HostTargetComm`, the host-port interface (HPI) is utilized instead of DMA. HPI is a 16-bit wide parallel port through which a PC host can directly access the C6x memory and peripherals. No specific configuration is required for the HPI communication on the DSP side, since the host program acts as master. However, the `intermediate` pointer should be passed to the host for having a proper memory access, because the triple buffering scheme keeps changing the address of the physical memory where the array of intermediate values is stored. Also, a variable should be used as a flag for data synchronization. A fixed physical memory address, say `0x00000200`, is allocated as a separate section, `.hpi`, to allow access from the host. This way the processed data can always be read by referring to its address stored in this section. Thus, the linker command file and the function `HostTargetComm` need to be changed as follows:

```
MEMORY
{
        VECS:  o = 00000000h    l = 00000200h
        HPI:   o = 00000200h    l = 00000008h
        IRAM:  o = 00000208h    l = 0000FDF8h
        CE0:   o = 80000000h    l = 01000000h
}

SECTIONS
{
    ...
```

```
        .hpi        >        HPI
    }


    #pragma DATA_SECTION(hpi_access, ".hpi")

    int hpi_access[2];    // pointer of intermediate data array

    ...

    void HostTargetComm(void)
    {
         hpi_access[0] = (int) intermediate;
         hpi_access[1] = 0xffffffff;         // Data is ready to be read

     // Wait until the flag is reset to 0 by Host after data transmission
          while(hpi_access[1] != 0x00000000);
    }
```

By using a function generator and an oscilloscope, the operation of the modified sampling program can be verified. When using EVM, to get data from the FIFO on the host side and display them on the PC monitor, a program called *Host.exe* running on the PC host is provided on the accompanying CD-ROM. This program utilizes the evm6x_read() API from the EVM host support library. When using DSK, the API function dsk6x_hpi_read() is used instead. This program, which is written in Microsoft™ Visual C++ using a Dialog wizard, starts a thread that continuously reads data from the pointer in the .hpi section and plots it on the monitor. The APIs used in the program are part of CCS.

The communication between the PC host and the DSP is achieved via RTDX for C6416/C6713. The accompanying CD-ROM contains the corresponding source codes with the RTDX APIs. A detailed description of RTDX appears in Chapter 10. The host application program for displaying and analyzing data via RTDX is also on the accompanying CD-ROM. Descriptions and examples of developing host application programs with Visual Basic, Visual C++, MATLAB, or Microsoft® Excel can be found at the TI website.

L6.1 DFT Implementation

DFT can be simply calculated from the equation

$$X[k] = \sum_{n=0}^{N-1} x[n] * W_N^{nk}, \quad k = 0,1,...,N-1, \tag{9.1}$$

where $W_N = e^{-j2\pi/N}$. This equation requires N complex multiplications and $N-1$ complex additions for each term. For all N terms, N^2 complex multiplications and $N^2 - N$ complex additions are needed. As is well known, this method is not efficient since the symmetry properties of the transform are not utilized. However, it is useful to implement the equation on the C6x as a comparison to the FFT implementation. The graphing capability of CCS is used here for this purpose. This is carried out in an offline manner because the amount of time required to do the DFT exceeds the duration of a frame capture.

First, a simple composite signal is generated in MATLAB with the frequency components at 750 Hz, 2500 Hz, and 3000 Hz. Saving two periods of this signal sampled at 8000 Hz results in a 64-point signal. Figure 9-7 shows the signal read into CCS and plotted using its graphing capability. The frequency content of the signal is also plotted based on a built-in FFT option.

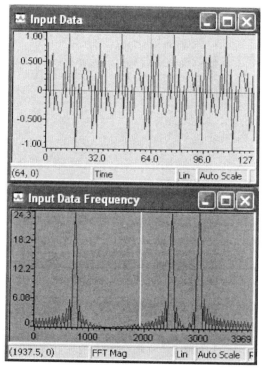

Figure 9-7: Input signal in time and frequency domains.

The DFT code used is the one appearing in the *TI Application Report SPRA291* [1]. Here is the code:

```
#include <math.h>
```

```
#include <math.h>
#include "params.h"

void dft(int N, COMPLEX *X){
    int n, k;
    double arg;
    int Xr[1024];
    int Xi[1024];
    short Wr, Wi;
    for(k=0; k<N; k++){
        Xr[k] = 0;
        Xi[k] = 0;
        for(n=0; n<N; n++){
            arg =(2*PI*k*n)/N;
            Wr = (short)((double)32767.0 * cos(arg));
            Wi = (short)((double)32767.0 * sin(arg));
            Xr[k] = Xr[k] + X[n].real * Wr + X[n].imag * Wi;
            Xi[k] = Xi[k] + X[n].imag * Wr - X[n].real * Wi;
        }
    }

    for (k=0;k<N;k++){
        X[k].real = (short)(Xr[k]>>15);
        X[k].imag = (short)(Xi[k]>>15);
    }
}
```

In order to use this code, the input has to be represented as complex numbers. This is done using a structure definition to create a complex variable with components real and imag. The main program used to perform DFT is as follows:

```
main()
{
    int i,j;
    COMPLEX x[128];
    int mag[128];

    /*Change input to Q-15*/
    for(i=0;i<128;i++)
    {
        x[i].real=0x7FFF * input_data[i];
        x[i].imag=0;
    }
    dft(128, x);

        for(i=0;i<128;i++)
            mag[i]=(x[i].real*x[i].real + x[i].imag*x[i].imag) << 1;
    return(0);
}
```

In this program, the input is converted to Q-15 format and stored in the complex structure, which is then used to call the DFT function. The magnitude of the DFT outcome is shown in Figure 9-8. As expected, there are three spikes, at 750Hz,

2500Hz, and 3000Hz. Notice that this code is quite inefficient, as it calculates each twiddle factor using the math library at every iteration. Running this code from the external SDRAM results in an execution time of about 1.6×10^9 cycles for a 128-point frame. As a result, the preceding DFT code cannot be run in real-time on the DSK, since only $18,750 \times 128 = 2.4 \times 10^6$ cycles are available to perform a 128-point transform.

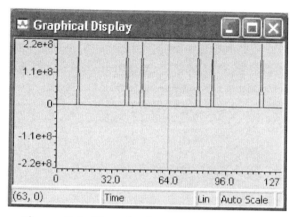

Figure 9-8: Magnitude response of DFT.

L6.2 FFT Implementation

To allow real-time processing, FFT is used which utilizes the symmetry properties of DFT. The approach of computing a $2N$-point FFT as mentioned in the *TI Application Report SPRA291* [1] is adopted here. This approach involves forming two new N-point signals $x_1[n]$ and $x_2[n]$ from the original $2N$-point signal $g[n]$ by splitting it into even and odd parts as follows:

$$x_1[n] = g[2n], \qquad 0 \le n \le N-1;$$
$$x_2[n] = g[2n+1]. \tag{9.2}$$

From the two sequences $x_1[n]$ and $x_2[n]$, a new complex sequence is defined as

$$x[n] = x_1[n] + jx_2[n], \qquad 0 \le n \le N-1. \tag{9.3}$$

To get $G[k]$, DFT of $g[n]$, the equation

$$G[k] = X[k]A[k] + X^*[N-k]B[k],$$
$$k = 0, 1, ..., N-1, \text{ with } X[N] = X[0],$$

(9.4)

is used, where

$$A[k] = \frac{1}{2}\left(1 - jW_{2N}^k\right)$$

(9.5)

and

$$B[k] = \frac{1}{2}\left(1 + jW_{2N}^k\right)$$

(9.6)

Only N points of $G[k]$ are computed from Eq. (9.4). The remaining points are found by using the complex conjugate property of $G[k]$, $G[2N-k] = G^*[k]$. As a result, a $2N$-point transform is calculated based on an N-point transform, leading to a reduction in the number of cycles. The codes for the functions (`split1`, `R4DigitRevIndexTableGen`, `digit_reverse`, and `radix4`) implementing this approach are provided in the *TI Application Report* [1].

Figure 9-9 shows the FFT outcome where the signal has been scaled down 0, 2, 4, and 5 times, respectively. The scaling is done to get rid of overflows, which are present for the scale factors 0, 2, and 4. As revealed by these figures, the input signal has to be scaled down five times to eliminate overflows. When the signal is scaled down five times, the expected peaks appear. The total number of cycles for this FFT is 56,383. Since this is less than the capture time available for a 128-point data frame at a sampling frequency of 8 kHz, it is expected that this algorithm would run in real-time on the DSK.

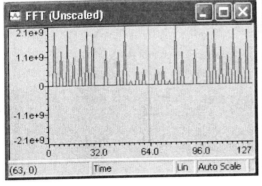
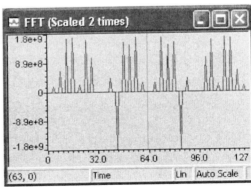

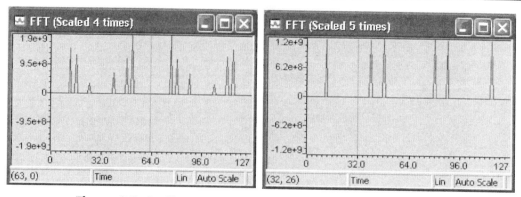

Figure 9-9: Scaling to get correct FFT magnitude response.

L6.3 Real-Time FFT

To perform FFT in real-time, the triple buffering program is used. A frame length of 128 is considered here. The output is observed by halting the processor through CCS. The animate feature of CCS cannot be used here since it slows down the processing and causes frames to overlap.

The following modifications are made to the triple buffering program to run the aforementioned FFT algorithm in real-time:

```
void wait_buffer(void)
{
        int n,k;
        short *p;

        while(index);

        p = input;
        input = output;
        output = intermediate;

        for (n=0; n<NUMPOINTS; n++)
        {
                x[n].imag = p[2*n + 1];        // x2(n) = g(2n + 1)
                x[n].real = p[2*n];            // x1(n) = g(2n)
        }

        radix4(NUMPOINTS, (short *)x, (short *)W4);
        digit_reverse((int *)x, IIndex, JIndex, count);
        x[NUMPOINTS].real = x[0].real;
        x[NUMPOINTS].imag = x[0].imag;

        split1(NUMPOINTS, x, A, B, G);
        G[NUMPOINTS].real = x[0].real - x[0].imag;
        G[NUMPOINTS].imag = 0;
```

```
for (k=1; k<NUMPOINTS; k++) {
        G[2*NUMPOINTS-k].real = G[k].real;
        G[2*NUMPOINTS-k].imag = -G[k].imag;
}
for (k=0; k<NUMDATA; k++) {
        mag1[k] = (G[k].real*G[k].real) << 1;
        mag2[k] = (G[k].imag*G[k].imag) << 1;
        mag[k] = mag1[k] + mag2[k];
}
intermediate = p;
HostTargetComm();
while(!index);
}
```

The `wait_buffer()` function is modified with the appropriate function calls so that when the `input` buffer is full, the transform is calculated and sent out to the host through the FIFO.

The functionality of the program can be verified by connecting a function generator to the line-in. The graphing capability of CCS can be used to plot the FFT outcome. By changing the frequency of the input, the spikes in the frequency response would move to left or right accordingly. Figure 9-10 illustrates the output for a 1 kHz and a 2 kHz sinusoidal signal. These snap shots are captured by halting the processor. The input here is scaled by shifting it right 4 bits.

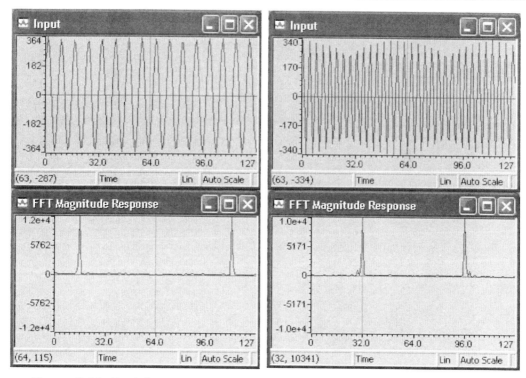

Figure 9-10: Real-time FFT magnitude response (a) *f* = 1 kHz (b) *f* = 2 kHz.

Bibliography

[1] Texas Instruments, *Application Report SPRA 291*, 1997.

CHAPTER 10

Real-Time Analysis and Scheduling

Figure 10-1 provides an overview of the conventional CCS debugging techniques based on breakpoints, probe points, and profiler. Although these debugging tools are very useful to see whether an application program is logically correct or not, when it comes to making sure that real-time deadlines are met, they have limitations. The so-called DSP/BIOS feature of CCS complements the traditional debugging techniques by providing mechanisms to analyze an application program as it runs on the target DSP without stopping the processor. In traditional debugging, the target DSP is normally stopped and a snap shot of the DSP state is examined. This is not an effective way to test for real-time glitches.

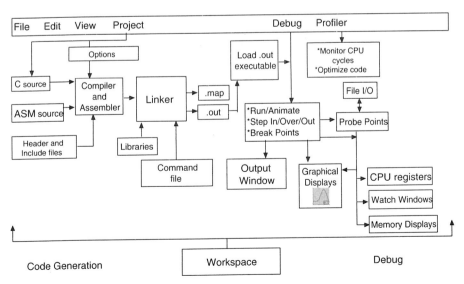

Figure 10-1: Code Composer Studio IDE.[†]

DSP/BIOS consists of a number of software modules that get glued to an application program to provide real-time analysis and scheduling capabilities. A listing of all available modules is provided in Figure 10-2. CCS provides an easy-to-use way to glue these modules to the application program. Figure 10-3 shows a sample C file with a BIOS object, a BIOS function, and its corresponding section names. The size of the DSP/BIOS portion of an application program is limited to a maximum of 2K words and is proportional to the number of modules and objects used.

Instrumentation/Real-Time Analysis	
LOG	Message log manager
STS	Statistics accumulator manager
TRC	Trace manager
RTDX	Real-Time DataeXchange Manager
Thread Types	
HWI	Hardware interrupt manager
SWI	Software interrupt manager
TSK	Multi-tasking manager
IDL	Idle function & process loop manager
Clock and Periodic Functions	
CLK	System clock manager
PRD	Periodic function manager

Comm/Synch Between Threads	
SEM	Semaphores manager
MBX	Mailboxes manager
LCK	Resource lock manager
Input/Output	
PIP	Data pipe manager
HST	Host Input/Output manager
SIO	Stream I/O manager
DEV	Device driver interface
Memory and Low-level Primitives	
MEM	Memory manager
SYS	System service manager
QUE	Queue manager
ATM	Atomic function
GBL	Global setting manager

Figure 10-2: DSP/BIOS API modules.[†]

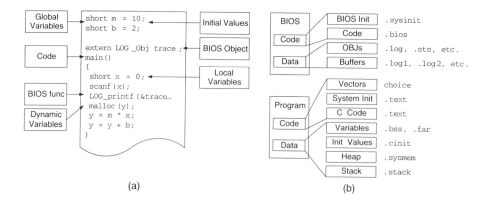

(a)   (b)

Figure 10-3: BIOS sections.[†]

Figure 10-4 shows a listing of the DSP/BIOS modules accessed by the **Configuration Tool** feature of CCS. This figure also shows the files generated by the **Configuration Tool**. The **Configuration Tool** is a visual editor which allows one to create module objects and set their properties. An application program can interact with objects by using DSP/BIOS API functions. In addition, DSP/BIOS plug-ins can be activated from the CCS environment, providing real-time instrumentation including log and statistics displays.

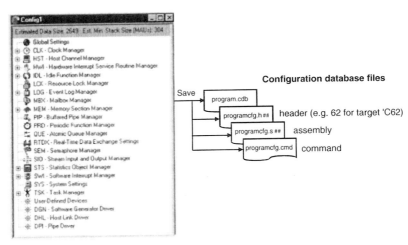

Figure 10-4: Files generated by the Configuration Tool.

Figure 10-5 illustrates all the files created within the CCS environment when using DSP/BIOS; the files with white background are created by the user and the ones with grey background by CCS. The naming convention used by the modules is shown in Figure 10-6. The datatypes associated with a module are defined in its header file. Header files of those modules which are used in real-time analysis of an application program must be included in the program.

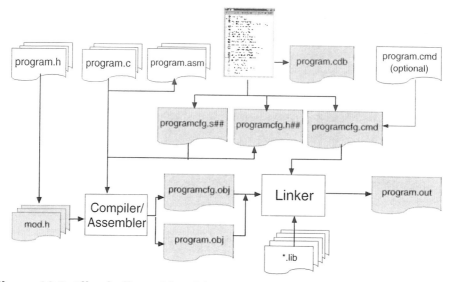

Figure 10-5: Files indicated in white created by user and in grey by CCS.†

CATEGORY	CONVENTION	EXAMPLE
Function Calls	MOD_lowercase	LOG_printf
Data Types	MOD_Titlecase	LOG_Obj
Constants	MOD_UPPERCASE	HWI_INT3
Internal Calls	MOD_F_lowercase	FXN_F_nop

Figure 10-6: Three letter prefix module naming conventions, capitalization convention distinguishes functions, types, and constants.

In essence, DSP/BIOS provides real-time analysis, real-time scheduling, and real-time data exchange capabilities for debugging application programs. As a result, one can make sure that an application program is meeting its real-time deadlines in addition to being logically correct.

10.1 Real-Time Analysis

There are two types of real-time constraints: (a) hard real-time, denoting critical real-time needs (i.e., timing needs that should be met to avoid system failure), and (b) soft real-time, denoting not-so-critical real-time needs that can be done as time becomes available. For example, as shown in Figure 10-7, the response to incoming samples must be done in a hard real-time manner in order not to lose any information, whereas data transfer from the target DSP to the host can be done in soft real-time.

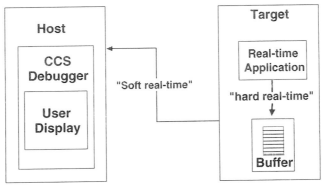

Figure 10-7: Hard and soft real-time: data buffered in hard real-time, and sent to host in soft real-time.

To monitor the status of a program in real-time, it is possible to use the C function printf () as defined in the real-time support library. However, this function takes too many cycles to run, an undesirable property as far as real-time performance is concerned. On the other hand, LOG_printf () is a LOG module API that creates a buffer. The buffer is then sent to the host in soft real-time. The buffer size *n* is specified as part of a LOG object. A LOG object can be configured in fixed or circular mode. The fixed mode captures the first *n* occurrences, while the circular mode captures the last *n* occurrences. LOG_printf () runs in much fewer clock cycles as compared with print (). Figures 10-8(a), 10-8(b), and 10-8(c) illustrate the difference between printf () and LOG_printf ().

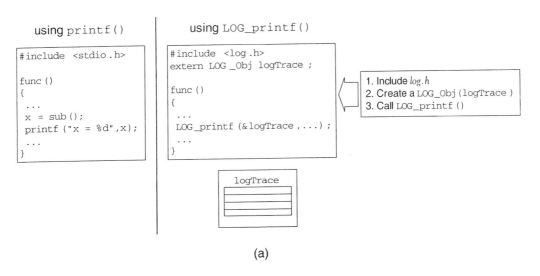

(a)

	printf()	LOG_printf()
Contained in	RTS Libraries	DSP/BIOS
Formats data with	Target	Host
Runs in the context of	Calling Function	Soft real-time
Written in	C	ASM
Optimization	No	Yes

(b)

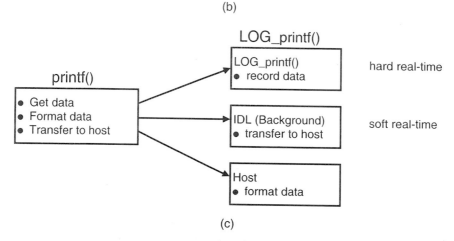

(c)

Figure 10-8: `printf()` vs. `LOG_printf()`.[†]

Statistics on a value can be captured by using the `STS_add()` API. Two other statistics API, `STS_set()` and `STS_delta()`, can be used to time a piece of code. The plug-in **Statistics View** window can be activated on the host as part of CCS to monitor the statistics associated with a variable. These statistics are reconfigured on the host as shown in Figure 10-9.

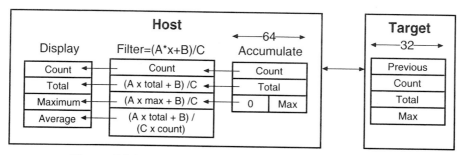

Figure 10-9: Reconfiguration of statistics on host.[†]

The trace module through the **RTA Control Panel** of DSP/BIOS allows various modules to be enabled or turned on so that only a specific or needed portion of the DSP/BIOS kernel is glued to the application program. The **RTA Control Panel** properties can be set to decide how often the host should poll the target DSP for various logging and statistics data. The **CPU Load Graph** is another plug-in instrumentation which shows the monitoring of the CPU active time as a program runs on the target DSP.

10.2 Real-Time Scheduling

This is done by breaking down the application program into threads each doing a specific function or task. Some of the threads may occur more often than others. Some of them may be subject to hard real-time and some to soft real-time constraints. The real-time need of the entire application program or system is met by appropriately prioritizing threads. This multithreaded real-time scheduling approach is what makes it possible to meet real-time timing deadlines.

Threads can be scheduled using hardware interrupts (ISRs) in a non-preemptive fashion by disabling hardware interrupts. The scheduling can be done in a preemptive fashion by prioritizing hardware interrupts. Considering that not all real-time situations can be handled by preemptive hardware interrupts, a more robust scheduling mechanism based on software interrupts is adopted in DSP/BIOS. Software interrupts automatically perform context switching (i.e., storage/retrieval of the DSP status to the time the interrupt occurred).

DSP/BIOS uses a background/foreground scheduling approach where the background consists of non-critical housekeeping threads such as transferring information to the host and instrumentation. These threads or functions are done in a round-robin fashion as part of idle or background loop IDL. The foreground consists of more critical threads. These threads are implemented via hardware (HWI module) and software (SWI module) interrupts. Hardware interrupts have higher priority than software interrupts. As indicated in Figure 10-10, normally software interrupts are used for deadlines of 100 microsec or more, and hardware interrupts for more restrictive deadlines of 2 microsec or more. The priority of software interrupts can easily be changed through SWI objects. Software interrupts can be posted unconditionally or conditionally via mailboxes. In essence, the DSP/BIOS scheduling is based on preemptive software interrupts. Figure 10-11 shows an example of how a single-thread ISR is converted to a hardware, software, and idle multithread program. In addition to hardware and software threads, the upgraded version of DSP/BIOS provides a task TSK module which can be used for posting threads capable of yielding to other threads.

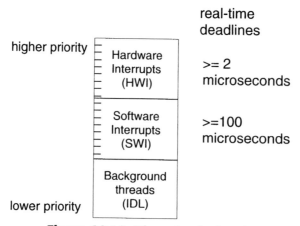

Figure 10-10: Thread priorites.[†]

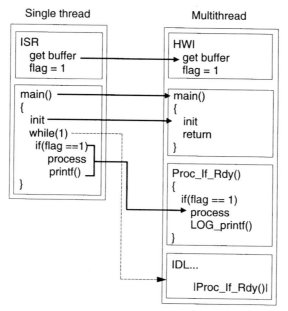

Figure 10-11: Multithread programming.[†]

There are basically two thread scheduling rules. The first rule is that if a higher priority thread becomes ready, the running thread is preempted. The second rule is that threads having the same priority are scheduled in a first-in first-out fashion. Three examples are shown in Figure 10-12 to illustrate how various threads run based on the scheduling rules. In this figure, a running thread is shown by shaded blocks and a thread in ready state by white blocks per time tick. The **Execution Graph** window feature of DSP/BIOS provides a visual display of execution of threads. It shows which thread is running and which threads are in ready state. It also provides useful feedback information on errors. Errors are generated when real-time deadlines are missed or when the system log is corrupted. An example of the **Execution Graph** window is shown in Figure 10-13. It should be noted that in the **Execution Graph** window, although time intervals between time ticks are the same, they may not be displayed as equal. This provides a more compact way to show all events between two successive time ticks.

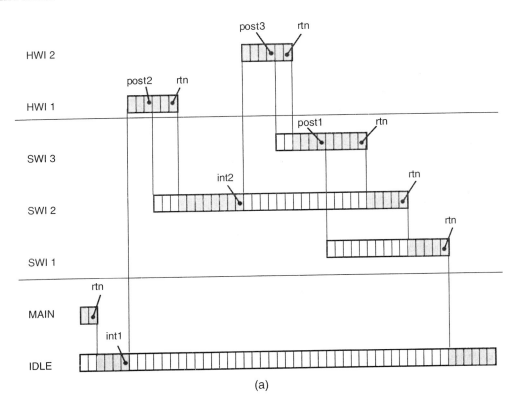

(a)

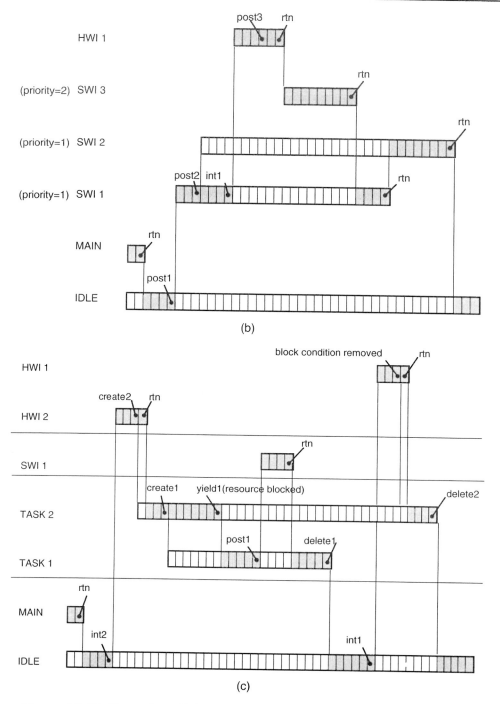

Figure 10-12: Examples of threads running based on scheduling rules, gray blocks indicate running status and white blocks ready status.[†]

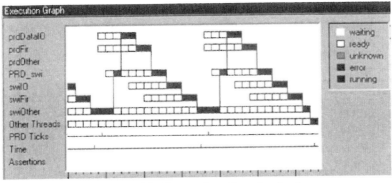

Figure 10-13: Execution Graph window.

There are two timers on the C6x, each is controlled by three memory-mapped registers: the Timer Control Register for setting the operating mode, the Timer Period Register for holding the number of clock cycles to count, and the Timer Counter Register for holding the current clock cycle count. As illustrated in Figure 10-14, the clock module CLK is used to set the on-chip timer registers for low (determined by the Timer Period Register) or high resolution (the CPU clock divided by 4) ticks. The clock APIs run as hardware functions. The PRD module is used to run threads that are to be executed periodically. The period is specified as part of a PRD object. The period APIs run as software functions. The CLK manager is used to drive the PRD module.

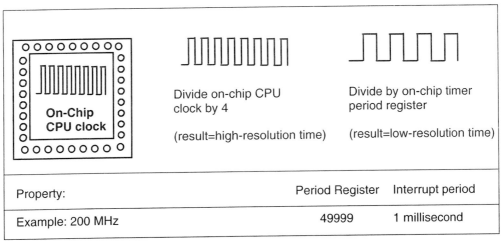

Figure 10-14: Low and high resolution clock ticks.[†]

Data frame synchronization and communication can be achieved by using the PIP module. A pipe consists of a specified number of frames having a specified size. It has two ends, a writer end and a reader end. The sequence of operations on the writer side consists of getting a free frame from the pipe via `PIP_alloc()`, writing to it, and putting it back in the pipe via `PIP_put()`, which runs the `notifyReader()` function. The sequence of operations on the reader side consists of getting a full frame from the pipe via `PIP_get()`, reading it, and putting the empty frame in the pipe via `PIP_free()`, which runs the `notifyWriter()` function. Figure 10-15 illustrates this process.

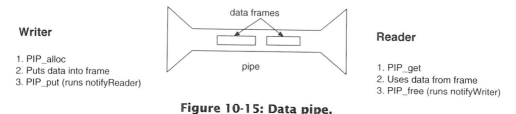

Writer

1. PIP_alloc
2. Puts data into frame
3. PIP_put (runs notifyReader)

Reader

1. PIP_get
2. Uses data from frame
3. PIP_free (runs notifyWriter)

Figure 10-15: Data pipe.

10.3 Real-Time Data Exchange

The RTDX (Real-Time Data Exchange) module can be used to exchange data between the DSP and the host without stopping the DSP. Similar to other modules, this exchange of information between the host and the DSP is done via the JTAG (Joint Test Action Group) connection, an industry-standard connection. The RTDX module provides a useful tool when values need to be modified on the fly as the DSP is running. As shown in Figure 10-16, RTDX consists of both target and host components, each running its own library. On the host side, various displays and analysis OLE (object linking and embedding) automation clients, such as LabVIEW, Visual Basic, Visual C++, can be used to display and send data to the application program. RTDX can be configured in two modes: non-continuous and continuous. In non-continuous mode, data is written to a log file on the host. This mode is normally used for recording purposes. In continuous mode, the data is buffered by the RTDX host library. This mode is normally used for continuously displaying data.

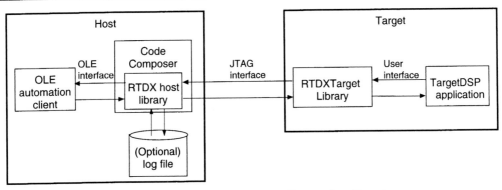

Figure 10-16: RTDX target/host dataflow.[†]

Labs 7 and 8 that follow provide a hands-on experience with the DSP/BIOS features of CCS. Lab 7 covers its real-time analysis and scheduling and Lab 8 its data synchronization and communication aspects. More details on the DSP/BIOS modules can be found in the *TMS320C6000 DSP/BIOS User's Guide* manual [1].

Bibliography

[1] Texas Instruments, *TMS320C6000 DSP/BIOS User's Guide*, Literature ID# SPRU 303B, 2000.

Lab 7: DSP/BIOS

The objective of this lab is to become familiar with the DSP/BIOS feature of CCS. DSP/BIOS includes a real-time library which allows one to interact with an application program in real-time as it runs on the target DSP. To build a program based on DSP/BIOS, the **Configuration Tool** feature of CCS needs to be used to create objects and set their properties. The **Configuration Tool** is opened by choosing the menu item **File → New → DSP/BIOS Configuration**. The configuration of a program can be saved via **File → Save**, and **Configuration Files (*.cdb)** in the **Save as type** drop-down box. A saved configuration file includes all the necessary files for generating an executable file.

A DSP/BIOS object can be created by right-clicking on a module displayed in the **Configuration Tool** window and by selecting **Insert**. For example, as shown in Figure 10-17, to create a PRD object, right-click on **PRD – Periodic Function Manager** and select **Insert PRD**. This adds a new object for the PRD module. An object can be renamed by right-clicking on its name and choosing **Rename** from the pop-up menu. Properties of an object can be displayed by right-clicking on the object icon and selecting **Properties** from the pop-up menu. From the property sheet, property settings can be readily changed.

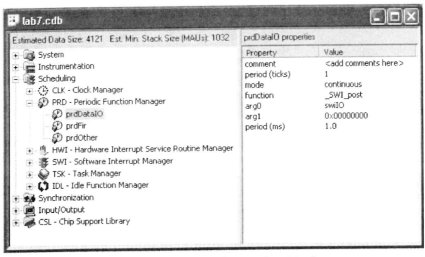

Figure 10-17: Configuration Tool.

L7.1 A DSP/BIOS-Based Program

The following C code is an example of a simple DSP/BIOS-based program. This program compares the performance of the DSP/BIOS API LOG_printf() to the function printf(), which is a part of the run-time support library:

```
#include <stdio.h>          // For printf();
#include <std.h>            // Header files needed for DSP BIOS
#include <log.h>
#include <sts.h>            // Header files added to support statistics
#include <clk.h>

void fun1();               // functions
void fun2();
extern LOG_Obj logTrace1;   // Objects created by the Configuration Tool
extern STS_Obj stsPrintf;
extern STS_Obj stsLogprintf;

void main()                // ======== main ========
{
    return;                // fall into DSP/BIOS idle loop
}

void fun1()
{
    static int i=0;
    i=CLK_gethtime();
    STS_set(&stsPrintf, CLK_gethtime());
    printf("loop: %d\n" , i);       // write a sting to stdio using printf
    STS_delta(&stsPrintf, CLK_gethtime());
    return;
}

void fun2()
{
    static int j=0;
    j=CLK_gethtime();
    STS_set(&stsLogprintf, CLK_gethtime());
    LOG_printf(&logTrace1, "loop: %d\n", j);
                           // write a string using BIOS LOG_printf object trace1
    STS_delta(&stsLogprintf, CLK_gethtime());
    return;
}
```

Two functions are declared in this program: fun1() and fun2(), one using printf() and the other LOG_printf(). These functions obtain the processing time and print them on the screen. Printing is the most widely used way to view results of a program. As shown in Figure 10-8, LOG_printf() is optimized to take much fewer instruction cycles than printf(). It sends buffered data to the host in soft real-time to avoid missing real-time deadlines. On the other hand, printf()

does not use the background scheduling approach for transferring data to the host and, hence, may cause the program to miss its real-time deadlines. At this point, it is worth mentioning that although it is possible to use **Watch Window**, this option interrupts the DSP in order to transfer data and does not meet the real-time requirement.

Note that appropriate header files should be included to build a DSP/BIOS-based program. Foremost, the header file `std.h` should be included whenever using any DSP/BIOS API. The header files, *log.h*, *sts.h*, and *clk.h*, corresponding to the three modules LOG, STS, and CLK, respectively, are included in the aforementioned program. Any created DSP/BIOS objects should also be declared. There are three declared objects here: `logTrace1`, `stsPrintf`, and `stsLogprintf`. The LOG object `logTrace1` managed by the LOG module allows real-time logging. The STS objects managed by the STS module store key statistics in real-time. The `STS_set()` and `STS_delta()` APIs use the information stored in the STS objects to compute the required number of instruction cycles to run `printf()` or `LOG_printf()`.

Now let us create the objects declared in the program. First, a configuration file is created by choosing **File** → **New** → **DSP/BIOS Configurations**. If a configuration file already exists, it can be activated by double-clicking on it in the **Project View** panel. To add a LOG object as part of the `LOG_printf()` API, right-click on **LOG-Event Log manager** in the Configuration Tool and select the option **Insert LOG** from the pop-up menu. This causes `LOG0` to be inserted. Since the name `logTrace1` is used here, rename this object by right-clicking on it and then by selecting **Rename**. Change the name to `logTrace1`. Right-click on `logTrace1` to change its properties. Select **Cancel** noting that the default settings are fine for this lab. Next, create two STS objects in a similar manner and rename them as `stsPrintf` and `stsLogprintf`. The properties of these objects will be discussed in the next section. Use **File** → **Save** to save the configuration file.

L7.2 DSP/BIOS Analysis and Instrumentation

Real-time analysis allows one to determine whether an application program is operating within its real-time deadlines and whether its timing can be improved. DSP/BIOS instrumentation APIs and DSP/BIOS plug-ins enable real-time data gathering and monitoring as an application program is running. For example, when using the instrumentation API `LOG_printf()`, the communication between the DSP and the host is performed during the idle state or in the background. The idle thread has the lowest priority. As a result, the real-time behavior of an application program is not affected.

In the preceding program, the instrumentation APIs STS_set() and STS_delta() are used to benchmark the functions printf() and LOG_printf(). STS_set() saves the value specified by CLK_gethtime() as the previous value in the STS object. STS_delta() subtracts this saved value from the value it is passed. Consequently, STS_delta() in conjunction with STS_set() provide the difference between the start and completion of the function in between. However, to obtain an accurate benchmarking outcome, the overhead associated with the instrumentation APIs should be subtracted. To calculate this overhead, the program should be run again by leaving out LOG_printf() and printf().

Before calculating the overhead, let us examine how the STS objects should be used during benchmarking. Since the STS objects count system ticks, they do not provide the actual CPU instruction cycles. A filtering operation on the host is normally performed to show the actual CPU instruction cycles. This is done by changing the properties of the STS objects via right-clicking and selecting **Properties** from the pop-up menu. In the properties box, go to the **unit type** field and choose **High resolution time based** from the drop-down menu. This changes the **host operation** field to **A*x** and the value in the **A** field to 4, as shown in Figure 10-18.

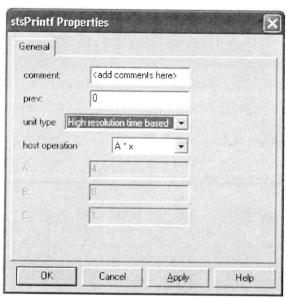

Figure 10-18: stsPrintf object properties.

One mechanism to come out of the idle state in `main()` is to use PRD objects to call `fun1()` and `fun2()`. In this lab, such an approach is adopted by activating PRD objects every 50msec. These objects are created by right-clicking on **PRD – Periodic Function Manager** and selecting **Insert PRD**. The objects need to be renamed as `prdPrintf` and `prdLogprintf`. For the object `prdPrintf`, change the properties as illustrated in Figure 10-19. Since the property **period(ticks)** is set to 50, this object calls the function mentioned in the property field **function** every 50 msec. This is because 1 tick (or timer interrupt) is set to 1000 microsec (or 1 msec) in the **CLK – Clock Manager** module. The property of this module is changed by right-clicking on **CLK – Clock Manager** and selecting **Property**, as shown in Figure 10-20. Notice that when specifying the function `fun1()` in the property field **function**, an underscore should be added before it. This rule holds for a C function to be run by DSP/BIOS objects. The underscore prefix is necessary because the **Configuration Tool** creates assembly source, and the C calling convention requires the underscore when calling C from assembly. For the `prdLogprintf` object, similarly enter `_fun2` in the property field **function**.

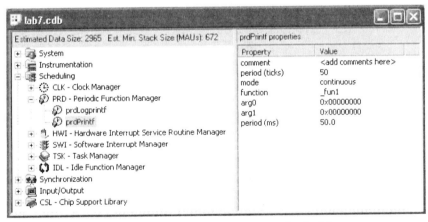

Figure 10-19: Property of prdPrintf object.

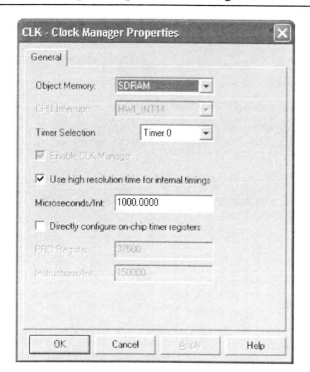

Figure 10-20:
Property of CLK module.

After building the program, in order to view the statistics information captured by the STS objects, choose the menu item **DSP/BIOS** → **Statistics View**. Then, right-click in this window and select **Property Page**. In the **Statistics View** Dialog Box, click on the objects stsPrintf and stsLogprintf, then click **OK**. You may wish to re-size the window so that all of the statistics can be viewed. Run the program. Without printf() or LOG_printf(), the average number of instruction cycles captured by the STS objects are 86 and 92, respectively, which correspond to the overhead for calling STS_set() and STS_delta(). Next, rebuild with printf() and LOG_printf(). In order to eliminate the overhead, change the properties of the STS objects as follows: **host operation = (A * x + B)**, **A** = 4 and **B** = –86 for stsPrintf and **B** = –92 for stsLogprintf. Run the program again. The **Statistics View** window displays 220790 instruction cycles for printf() and 56 for LOG_printf(), as shown in Figure 10-21.

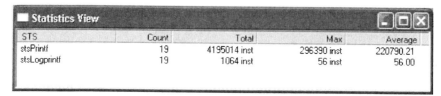

STS	Count	Total	Max	Average
stsPrintf	19	4195014 inst	296390 inst	220790.21
stsLogprintf	19	1064 inst	56 inst	56.00

Figure 10-21: Statistics View.

The output of LOG_printf() can be seen via a message log window. Select the menu item **DSP/BIOS → Message Log**. A new window will appear. This window should then be linked to the LOG object. In the drop-down box **Log Name** of the **Message Log** window, select logTrace1.

L7.3 Multithread Scheduling

Real-time scheduling involves breaking a program into multiple threads in order to meet a specified real-time throughput. The Lab 4 program is used here to study the real-time scheduling issues. First, the ISR in Lab 4 is modified as follows:

```
int cnt=0;

interrupt void serialPortRcvISR(void)
{
    ...
    if(cnt++ == 3)
    {
        otherProcessing(400);
        cnt = 0;
    }
    ...
}
```

The function otherProcessing() does no specific processing and merely consumes CPU time. This function is shown next:

```
        .def _otherProcessing
        .sect ".otherProcessing"
    N   .set    1000

;       void otherProcessing(int loopCount)
_otherProcessing:
        mv a4, b0               ; use b0 as loop counter
        mvk N,b1
        mpy b1,b0,b0
        nop
        shru b0,3,b0            ; (loop counter)= (# loops)/8
loop:
        sub b0,1,b0
        nop
 [b0] b loop
        nop 5
        b b3
        nop 5                   ; return
        .end
```

This function sets up a counter using the value passed to it. Let this value be 400. The counter is decreased one at a time in a loop, thus consuming CPU time. This function, of course, can be replaced with an actual processing code.

After building the project, connect the function generator and oscilloscope to the DSK board. Then, run the program. It is observed that the ISR does not meet real-time deadlines due to the extra processing required by the function `otherProcessing()`. In other words, the ISR misses input samples and fails to produce the desired output signal.

Now, let us perform real-time scheduling by breaking up the ISR into three functions or threads, `dataIO()`, `fir()`, and `otherProcessing()`, as follows:

```
/* Include header files here */
...

#define N 10
/* Declare global variables here */
...

Void main()
{
    index = 0;
    return;                            /* fall into DSP/BIOS idle loop */
}

Void dataIO()
{
    temp = inputBuffer[index++];       /* read a sample */
    index = (index == N) ? 0: index;   /* Restart */
}

void fir()
{
    int i,result;
    result = 0;

    for(i=N-1;i>=0;i--)                /* Update array samples */
        samples[i+1] = samples[i];
    samples[0] = temp;

    for(i=0;i<=N;i++)                  /* Filtering */
        result += ( _mpyhl(samples[i],b[i]) ) << 1;
}
```

The thread dataIO() reads one sample from inputBuffer whenever it is called. This thread simulates the operation of the MCBSP_read() API. The thread fir() performs FIR filtering. Let us now use three PRD objects to run these threads. Since otherProcessing() is called after every three input samples, it is not necessary to run all three threads or functions at the same period tick. The prdDataIO object runs the function dataIO(), and prdFir object runs the function fir() every 1 msec. The prdOther object runs the function otherProcessing() every 4 msec. The property settings of these PRD objects are shown in Figure 10-22.

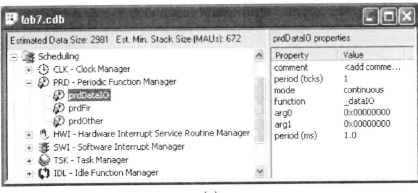

(a)

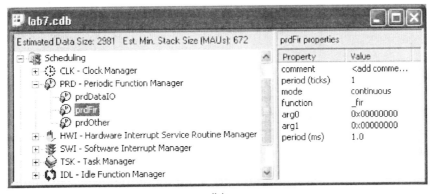

(b)

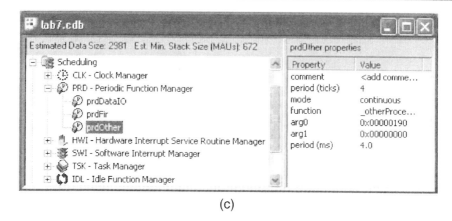

(c)

Figure 10-22: PRD objects for real-time analysis.

After building the project, to see whether the threads meet their real-time deadlines, choose **DSP/BIOS → RTA Control Panel** and place check marks in the boxes as indicated in Figure 10-23. Also, enable the global tracing option. Then, invoke the **Execution Graph** by choosing **DSP/BIOS → Execution Graph**. Right-click on the **RTA Control Panel** and choose **Property Page** from the pop-up menu. Run the program. The Execution Graph should look like the one shown in Figure 10-24.

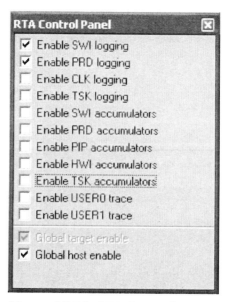

Figure 10-23: RTA Control Panel.

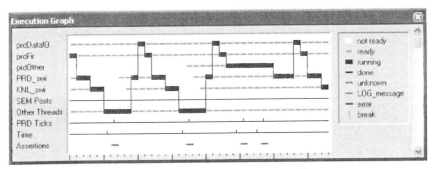

Figure 10-24: Execution Graph.

Any missed deadline error appears in the **Assertions** row of the **Execution Graph**. From Figure 10-24, it can be seen that there are such errors in the preceding multithread program. Another way to see the same information is via the message log window **Execution Graph Details**, by choosing **DSP/BIOS → Message Log**. In the **Log Name** field of the window, choose **Execution Graph Details** and click **OK**. An **Execution Graph Details** window should appear as shown in Figure 10-25. The information in this window indicates that prdFir() is missing its real-time deadlines. Figure 10-26 shows the CPU load when the program is running. To invoke this window, choose the menu item **DSP/BIOS → CPU Load Graph**.

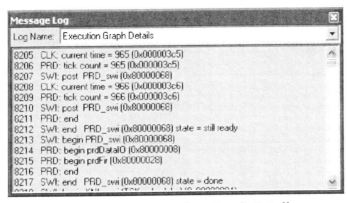

Figure 10-25: Execution Graph Details.

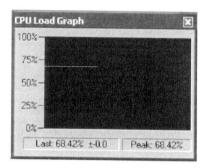

Figure 10-26: CPU Load Graph.

To overcome real-time errors, the scheduling of threads need to be changed by assigning different priorities to them. As shown in Figure 10-25, prdFir() is missing its real-time deadlines. This is due to the fact that periodic functions execute at the same priority level, since they run as part of the same software interrupt PRD_swi. This scheduling problem is overcome by allowing each periodic object to post a software interrupt (SWI) object, which then calls the appropriate thread or function.

A SWI object has five properties: **priority, function, arg0, arg1**, and **mailbox**. The property **function** causes a specified function to be called when the SWI object is posted. The arguments **arg0** and **arg1** are passed to the function. The property **priority** stores the priority level assigned to the SWI object. The **mailbox** property will be covered in Lab 8. In this lab, three SWI objects are created: swiIO, swiFir, and swiOther. Instead of the PRD objects, the SWI objects are used to run the threads dataIO(), fir(), and otherProcessing(). The swiIO object runs dataIO(), the swiFir object runs fir(), and the swiOther object runs otherProcessing(). Assuming that the real-time constraint of otherProcessing() is not as demanding as dataIO(), the priority of swiIO is set to 3 and the priority of swiOther to 1. The property settings of the SWI objects are shown in Figure 10-27.

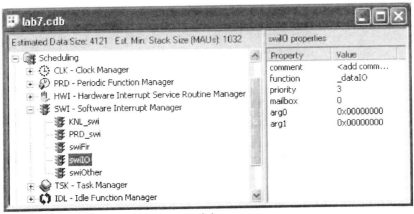

(a)

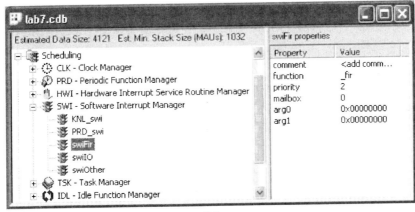

(b)

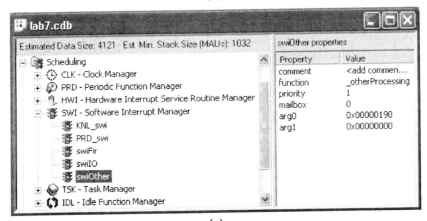

(c)

Figure 10-27: Properties of SWI objects.

Now that the SWI objects are ready to call the threads or functions, three PRD objects need to be set up to post the software interrupts. This is achieved by changing the properties of the original PRD objects, as shown in Figure 10-28. The PRD_swi thread runs the PRD functions associated with prdDataIO and prdFir every 1 msec and those associated with prdOther every 4 msec. In other words, the PRD functions post the software interrupts associated with the SWI objects. For instance, the prdDataIO object runs the function SWI_post(swiIO), which posts the software interrupt that in turn runs the function dataIO(). Although the software interrupts for both swiIO and swiFir are posted every 1 msec, the function da-taIO() runs first because the associated swiIO has a higher priority than swiFir. After the dataIO() function completes, the fir() function runs. The software interrupt for swiOther is posted every 4 msec, causing its associated function, otherProcession(), to run. However, when a higher priority thread becomes ready, otherProcessing() is preempted. Figure 10-29 shows the scheduling of the periodic and software threads.

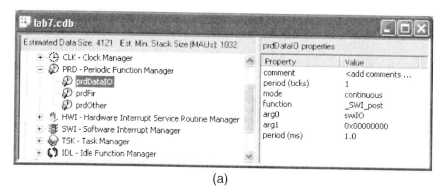

(a)

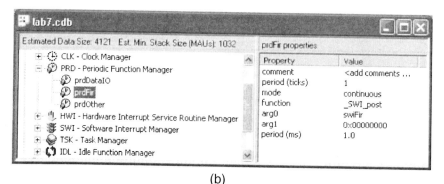

(b)

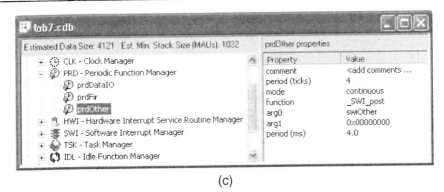

(c)

Figure 10-28: Properties of PRD objects for real-time scheduling.

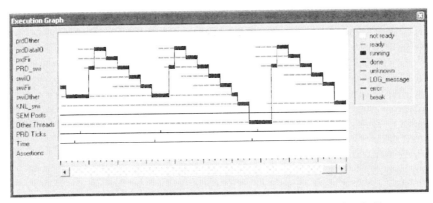

Figure 10-29: Execution Graph after real-time scheduling.

As illustrated in Figure 10-29, no real-time error is observed in the **Assertions** row because the threads are scheduled in such a way that they all meet their real-time deadlines. In general, critical and frequent events such as sampling should be assigned a higher priority. Next, let us go back to the ISR in the FIR filtering program. Based on the aforementioned real-time scheduling, the otherProcessing() thread can be moved into the while loop in main(), since the while loop runs on a lower priority (or in the background). The following piece of code shows this modification:

```
...
main()
{
    ...
    while (1)
    {
        if(cnt == 0)
            otherProcessing(400);
    }
}

interrupt void serialPortRcvISR(void)
{
    ...
    if( cnt++ == 3)    // Run every four samples.
    {
        cnt=0;
    }

    result = result >> 16;
    MCBSP_write(hMcbsp, result);
}
```

It can be observed that this program yields the correct FIR filtering output by connecting a function generator and an oscilloscope to the DSK board.

Lab 8: Data Synchronization and Communication

The objective of this lab is to become familiar with the data synchronization and communication capabilities of DSP/BIOS by using the FFT program discussed in Lab 6. This program uses hardware interrupts to read and process input samples entering the serial port 1 of the DSK, or serial port 0 of the EVM.

In the DSP/BIOS-based version of this program, the hardware module HWI is used to manage hardware interrupts. Let us use the interrupt INT11 to read input samples from the serial port 1 considering the DSK target. This requires that the properties of the HWI_INT11 object are appropriately set. First, create a new configuration by choosing **File → New → DSP/BIOS Configuration**. Then, click the + sign next to the **HWI – Hardware Interrupt Service Routine manager** menu under **Instrumentation** category, and right-click on **HWI_INT11** and select **Properties** from the pop-up menu. The window **HWI INT11 Properties** will appear. Select MCSP_1_Receive in the property field **interrupt source** so that the multichannel buffered serial port 1 is used as the interrupt source. Next, specify the function by entering _codec_isr in the property field **function**. This function reads a 32-bit input sample from the DRR and stores it into the location indicated by the global pointer rxPtr, which is explained shortly. The source codes of codec_isr() are provided on the accompanying CD-ROM.

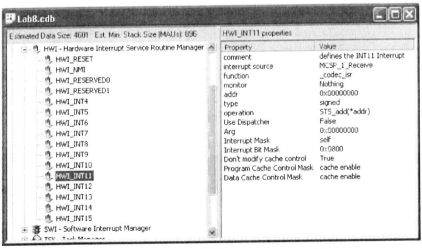

Figure 10-30: HWI object property setting.

Figure 10-30 shows the properties of the HWI_INT11 object. This object is configured to run codec_isr() whenever the hardware interrupt occurs.

The FFT program processes a frame of data when a specified number of samples are collected into an input frame buffer. This program also uses an output frame buffer to send out processed data frames. Since the FFT function should be performed only when the input frame is full and the output frame is empty, the execution of the FFT function needs to be synchronized with the status of the frames. In other words, the FFT function should be aware of whether or not the input frame is full and the output frame is empty. The mailbox property of SWI objects is used here for this purpose.

The mailbox property provides a conditional posting of software interrupts. Only when the mailbox in a SWI object becomes zero is the software interrupt posted by the SWI object. In this lab, a SWI object swiFFT is created and configured to run the FFT function when the mailbox value becomes zero. Initially, the mailbox is set to have a nonzero value of 3 (or 11 in binary). In order to run the FFT function, all the bits in the mailbox should be reset to zero. One possibility is to reset bit 0 to zero when the output frame is empty and reset bit 1 to zero when the input frame is full. The swiFFT object is therefore configured as shown in Figure 10-31. In order to create this object, right-click on **SWI - Software Interrupt Manager** under the **Scheduling** category inside the **Configuration Tool** window and select **Insert SWI** from the pop-up menu. A SWI0 object will be generated. Rename it swiFFT. Change the properties by right-clicking on the swiFFT object and selecting **Properties**. Inside the dialog box, change the properties as shown in Figure 10-31. The fft() function is assigned to the property **function** so that it is executed when the mailbox value becomes zero.

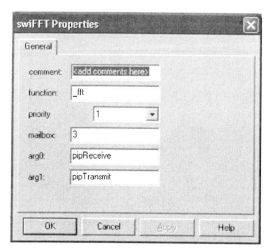

**Figure 10-31:
Properties of swiFFT object.**

The property settings in Figure 10-31 show that the function `fft()` takes two arguments: `pipReceive` and `pipTransmit`. These are PIP objects, which are used as the input frame and the output frame for the `fft()` function. The PIP module manages these frames. The interrupt service routine `codec_isr()` copies data from the DRR to the input frame via the `pipRecevice` object and from the output frame to the DXR via the `pipTransmit` object. The PIP objects need to be created and configured so that they can reset appropriate bits in the `swiFFT`'s mailbox to zero, causing the `fft()` function to run with a full input frame and an empty output frame. To create the `pipReceive` object, right-click on **PIP – Buffered Pipe Manager** under **Input/Output** category inside the **Configuration Tool** window, and choose **Insert PIP** from the pop-up menu. A `PIP0` object will be created. Rename it `pipReceive` by right-clicking on it and selecting **Rename**. The `pipTransmit` object is created in the same manner. The properties are then modified to meet the synchronization between the `swiFFT` and PIP objects. Figure 10-32 shows the properties of the `pipReceive` object, which is configured to clear bit 1 in the `swiFFT`'s mailbox when the input frame is full and ready to be processed. To change the properties of the `pipReceive` object, right-click on it and select **Properties** from the pop-up menu. A dialog box will appear in which new property values can be entered. The properties of the `pipTransmit` object are changed in a similar manner.

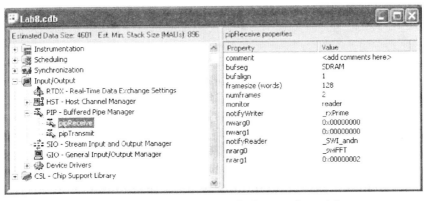

Figure 10-32: Properties of pipReceive object.

As shown in Figure 10-32, the function `_SWI_andn` is specified in the property field **notifyReader**. This property assigns the function to run when the input frame buffer is full and ready to be processed. As a result, whenever the input frame is full, the `_SWI_andn` function is executed. The `_SWI_andn` function clears the bits in the mailbox and posts a software interrupt. Its first argument, **nrarg0**, specifies the SWI object to be applied, and its second argument, **nrarg1**, denotes the mask. The mailbox value is reset by the bitwise logical AND NOT operator: `mailbox = mailbox AND (NOT mask)`. Because the `pipReceive` object has the mask value of 2 (or 10 in binary), it resets bit 1 in the mailbox to zero. Consequently, since `swiFFT` is the first argument of `_SWI_andn`, the `pipReceive` object resets bit 1 of the `swiFFT`'s mailbox whenever the input frame is full and ready to be processed.

Now that bit 1 of the `swiFFT`'s mailbox is reset to zero by the `pipReceive` object, the only condition to run the FFT function is to reset bit 0 considering that 3 (or 11) was the initial value in the mailbox. The `pipTransmit` object completes the synchronization process. Figure 10-33 shows the properties of the `pipTransmit` object, which is configured to clear bit 0 of the `swiFFT`'s mailbox when an empty frame is available. Note that the property field `notifyWriter` is set to `_SWI_andn`. This property specifies the function to run when an empty frame is available. Under such a condition, the `_SWI_andn` function runs with the mask value of 1 (or 01 in binary) and resets bit 0 of the `swiFFT`'s mailbox to zero. Hence, when the input frame is full and the output frame is empty, `pipReceive` and `pipTransmit` reset the `swiFFT`'s mailbox to zero, causing the FFT function to run.

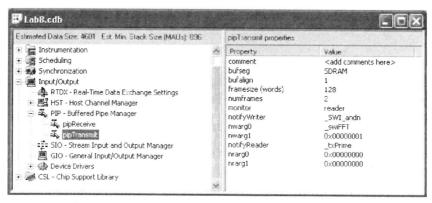

Figure 10-33: Properties of pipTransmit object.

Next, let us see how the FFT function makes use of data frames in the PIP objects. The following piece of code shows the sequence of events:

```
Void fft(PIP_Obj *in, PIP_Obj *out)
{
    int          *src, *dst;
    ...

    PIP_get(in);
    PIP_alloc(out);
    src = PIP_getReaderAddr(in);
    dst = PIP_getWriterAddr(out);
    size = PIP_getReaderSize(in);
    PIP_setWriterSize(out,size);

    /* FFT */
    for (n=0; n<NUMPOINTS; n++)
    {
      x[n].imag = src[2*n + 1];
      x[n].real = src[2*n];

      ...
    }

    ...
    for (; size > 0; size--)        //---------Copy input data into output
        *dst++ = *src++;
    PIP_put(out);
    PIP_free(in);
}
```

The first argument of the FFT function, `in`, is the `pipReceive` object and the second argument, `out`, is the `pipTransmit` object, as indicated in Figure 10-31. In order to use the frames in the PIP objects, first the `PIP_get()` and `PIP_alloc()` API functions should be called. The `PIP_get()` API gets a full frame from the `pipRecevie` object and the `PIP_alloc()` API allocates an empty frame from the `pipTransmit` object. Normally, the `PIP_get()` API is followed by the `PIP_getReaderAddr()` API, which returns the address for the reading process. Similarly, the `PIP_alloc()` API is followed by the `PIP_getWriterAddr()` API, which returns the address for the writing process. A pointer `src` is therefore used to read from the input frame and a pointer `dst` to write to the output frame. In this program, the FFT function processes the data stored in `src`. Before leaving the `fft()` function, the `PIP_put()` API is called to put the full frame into the `pipTransmit` object. Normally, this API is used together with the `PIP_alloc()` API because `PIP_put()` puts a frame allocated by `PIP_alloc()` into a PIP object after the frame is full. Similarly, the `PIP_free()` API is used together with the `PIP_get()` API because `PIP_free()` releases the frame for `PIP_get()` after it is read. The released frame is recycled so that it can be reused by the `PIP_alloc()` API.

Note that the value of the property **notifyWriter** in the `pipReceive` object is set to _rxPrime, as shown in Figure 10-32. Therefore, the function `rxPrime()` is called when a frame of free space is available in the `pipReceive` object. The following piece of code shows the relevant part in the `rxPrime()` function:

```
void rxPrime(void)
{
    PIP_Obj     *rxPipe = &pipReceive;
    ...
    if (rxCount == 0 && PIP_getWriterNumFrames(rxPipe) > 0) {
        PIP_alloc(rxPipe);
        rxPtr = PIP_getWriterAddr(rxPipe);
        rxCount = PIP_getWriterSize(rxPipe);
    }
    ...
}
```

The global variable `rxCount` keeps track of the remaining number of words for filling up the current `rxPipe` (or `pipReceive`) frame. In the `codec_isr()` function, `rxCount` is decreased by one whenever a sample is read from the DRR and copied into the `rxPipe` frame. When this frame becomes full and ready to be put into `rxPipe`, `rxCount` becomes zero. Then, this function allocates the next empty frame from `rxPipe` by calling `PIP_alloc(rxPipe)`. The address of the frame is set to the global variable `rxPtr` so that `codec_isr()` can copy the content of the DRR into `rxPtr` by calling `PIP_getWriterAddr(rxPipe)`. In the `codec_isr()` function, `rxPtr` is increased by one to point to the next location whenever the DRR content is copied.

As shown in Figure 10-33, _txPrime is written in the property field **notifyReader**. Therefore, the function `txPrime()` runs when a frame is full and ready to be used. The following piece of code shows the relevant part in the `txPrime()` function:

```
void txPrime(void)
{
    PIP_Obj     *txPipe = &pipTransmit;
    ...
    if (txCount == 0 && PIP_getReaderNumFrames(txPipe) > 0) {
        PIP_get(txPipe);
        txPtr = PIP_getReaderAddr(txPipe);
        txCount = PIP_getReaderSize(txPipe);
    }
    ...
}
```

The global variable `txCount` keeps track of the remaining number of words for transmitting the current `txPipe` (or `pipTransmit`) frame. In the function `codec_isr()`, `txCount` is decreased by one whenever a sample is copied from the `txPipe` frame and written to the DXR. When all the samples in this frame are written, `txCount` becomes zero. Then, this function gets the next full frame from `txPipe` by calling `PIP_get(txPipe)`. The address of the frame is set to the global variable `txPtr` by calling `PIP_getReaderAddr(txPipe)` so that `codec_isr()` can copy the content of `txPtr` into the DRR. The pointer `txPtr` is increased by one to point to the next location whenever a sample is written to the DXR.

After properly configuring the PIP and SWI objects, the program is built. The entire DSP/BIOS version of the FFT files are provided on the accpompanying CD-ROM. In order to verify the operation of the DSP/BIOS-based FFT program, connect a function generator to the DSK board and run the program. Figure 10-34 shows a snapshot of the CCS animation feature. This is done by setting a breakpoint at the end of the FFT function and by opening a graphical display window via the menu item **View →
Graph → Time/Frequency**. Place the global variable mag in the field **Start Address** to display the FFT magnitude values. Then, select the menu item **Debug → Animate** to start the animation. As the input frequency from the function generator is changed, the peaks in the graphical display window should move accordingly.

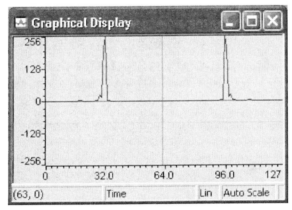

Figure 10-34: FFT magnitude.

The FFT magnitude can be sent to the PC host by using the DMA as discussed in Lab 6.

When using DSK, the host-target data transfer is achieved by RTDX via the JTAG interface. Since the magnitude of the FFT is always stored into the variable mag, there is no need to deploy circular buffering as done in Lab 6. To configure RTDX, a RTDX object for output, ochan, is inserted at **Input/Output** category of the DSP/BIOS configuration. After the initialization via the RTDX_enableOutput() API, the RTDX_write() API is used to write a new mag to the output channel.

As shown next, the function HostTargetComm() can be used to perform the same operation when using EVM:

```
void HostTargetComm()
{
    dma_reset();
    dma_init(    2,                        //Channel
                 0x0A000010u,              //Primary Control Register
                 0x0000000Au,              //Secondary Control Register
                 (unsigned int) mag,       //Source Address
                 0x01710000u,              //Destination Address
                 0x00010080u);             //Transfer Counter Register

    DMA_START(2);
}
```

The DMA uses the global variable mag as the source address and the memory location 0x0171000, which is dedicated for FIFO access in PCI bus transfers, as the destination address. The program *Host.exe*, provided on the accompanying CD-ROM, is written to display the FFT magnitude on the PC monitor. This program makes use of the EVM API evm6x_read(). This API transfers data from the DSP to the host. In this lab, a PRD object prdComm is created and configured to run the host-target communication function HostTargetComm() every 8 msec, as shown in Figure 10-35. After adding the HostTargetComm() function to the FFT program, build the program and run it. To observe the FFT magnitude in real-time, also run *Host.exe*. The **CPU load Graph** plug-in can be used to verify that the DMA transfers the contents of mag independently of the CPU. As shown in Figure 10-36, the CPU load remains almost the same while the DMA transfer is running. To invoke the **CPU load Graph** plug-in, choose the menu item **DSP/BIOS → CPU Load Graph**.

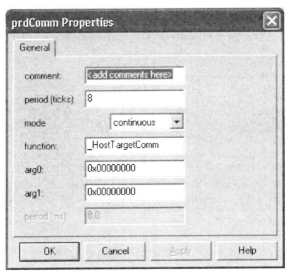

Figure 10-35: Properties of prdComm.

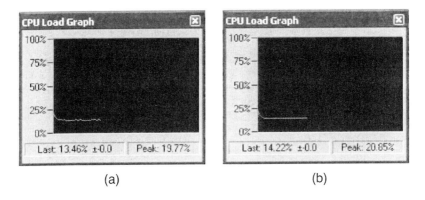

(a)   (b)

**Figure 10-36: CPU Load Graph; (a) CPU load before DMA transfer,
b) CPU load while DMA transfer is running.**

L8.1 Prioritization of Threads

Instead of using the function generator to generate input samples, a CD player can be connected to the input jack of the audio daughter card. Of course, a pair of amplified speakers should be connected to the output jack of the daughter card to hear the sound. Now, let us examine the effect on the sound quality by changing the CPU load. To change the CPU load, a PRD object prdLoad is created and configured to run the function changeload() every 8 msec, as illustrated in Figure 10-37. The function changeload() calls the otherProcessing() function. The CPU

261

load is determined by the global variable `loadVal`, which is passed to the `other-Processing()` function. The function `changeload()` is shown next:

```
Void changeload(Int prd_ms)
{
   ...
   if (loadVal)
       otherProcessing(loadVal);
}
```

The global variable `loadVal` can be set by choosing **Edit → Edit Variable** to invoke the Edit Variable dialog box. In this dialog box, write `loadVal` in the field `variable` and the desired number in the field `value`. The `prdLoad` object will run the `changeload()` function every 8 msec.

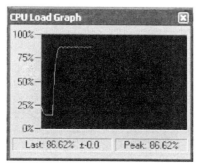

Figure 10-37: CPU load with loadVal = 900.

To observe the impact of the CPU load, let us build the program and run it while playing a CD. When the `loadVal` is changed to 900, the CPU load increases to about 87%, as shown in Figure 10-37, and the sound quality is degraded. The reason for this degradation can be seen by activating the Execution Graph Details window. As shown in Figure 10-38(a), when `loadVal` is zero, the `swiFFT` and `PRD_swi` threads complete their tasks without any problem. However, when `loadVal` is 900, these threads cannot complete their tasks and frequently go into the ready state, as shown in Figure 10-38(b). The **Execution Graph** in Figure 10-39 provides a graphical display of this situation. Since the `swiFFT` thread is competing with the `PRD_swi` thread, for large load values, the `fft()` function sits waiting to be executed. The `PRD_swi` thread executes all the PRD objects, so it eventually runs the CPU load function `otherProcessing()`. Consequently, since the audio from the CD player is copied into the output frame by the `fft()` function as part of the `swiFFT` thread, the sound quality suffers.

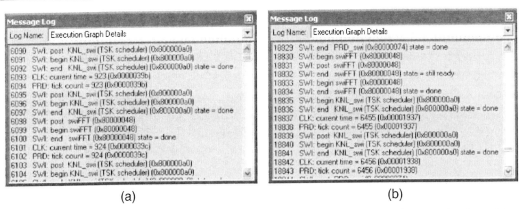

Figure 10-38: Execution Graph Details: (a) for loadVal = 0, (b) for loadVal = 900.

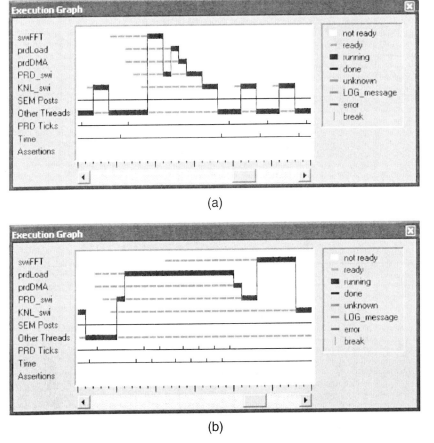

(a)

(b)

Figure 10-39: Execution Graph: (a) for loadVal = 0, (b) for loadVal = 900.

To solve this problem, the threads need to be properly prioritized. Let us assign a higher priority to the swiFFT thread. One simple way to do this is via the drag-and-drop method. Click on **SWI – Software Interrupt Manager** under **Scheduling** category in the **Configuration Tool**. On the right side of the window, click and hold the left mouse button on the PRD_swi icon, drag it to Priority 1, and release the button to drop it. This way the priority of the PRD_swi thread becomes 1. Similarly, move the swiFFT icon to Priority 2 so that it gets a higher priority than PRD_swi. Now build the program and run it. The sound quality remains unaffected even though loadVal = 900. The DSP/BIOS plug-ins in Figure 10-40 illustrate that the swiFFT thread no longer waits to be executed. Notice that the CPU load is now about 87%.

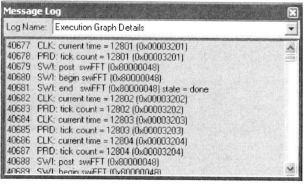

(a)

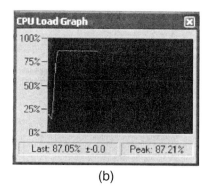

(b)

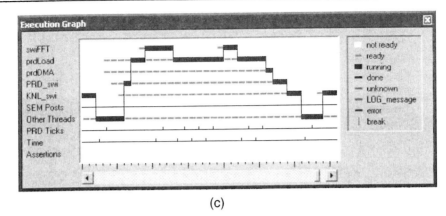

(c)

**Figure 10-40: DSP/BIOS plug-ins with loadVal = 900 after the prioritization:
(a) Execution Graph Details, (b) CPU Load, and (c) Execution Graph.**

L8.2 RTDX

The CPU load can be changed by using the RTDX module. This module allows loadVal to be transferred from the host to the DSP while the program is running. The following piece of code shows the parts to be added to the original program to access the RTDX module (the entire program is provided on the accompanying CD):

```
#include <rtdx.h>

RTDX_CreateInputChannel(writeload);
RTDX_CreateOutputChannel(readload);

main()
{
    ...
    RTDX_enableInput(&writeload);
    RTDX_enableOutput(&readload);
    return;
}

Void changeload(Int prd_ms)
{
    ...
    if (!RTDX_channelBusy(&writeload)) {
                        // Reads new loadVal sent from the PC host.
        RTDX_readNB(&writeload, &loadVal, sizeof(loadVal));

        if (oldLoad != loadVal ) {
            oldLoad = loadVal;
            RTDX_write(&readload, &loadVal, sizeof(loadVal));
            LOG_printf(&logTrace, "CPU load: new load = %d000
instructions every %d ms", loadVal, prd_ms);
        }
```

```
    }

    if (loadVal)
        otherProcessing(loadVal);
}
```

In order to use the indicated RTDX APIs, the program should include the header file *rtdx.h*. The RTDX input channel structure is declared and initialized by the macro `RTDX_CreateInputChannel`. Since it is declared as a global variable, it can be accessed anywhere in the program. Similarly, the `RTDX_CreateOutputChannel` macro defines and initializes the RTDX output channel structure. Because these channels are disabled during the initialization, they need to be enabled in `main()` by using the `RTDX_enableInput()` and `RTDX_enableOutput()` APIs. The input channel is examined by the `RTDX_channelBusy()` API to see whether it is busy or not. If it is not busy, data is read from the input channel by using the `RTDX_readNB()` API. This API posts a request to the RTDX host library that the DSP application program is ready to receive data. The DSP program keeps running at this point. When the `RTDX_read()` API is used, the DSP program stops until it receives data from the input channel. The `RTDX_write()` API is used to write a new `loadVal` to the output channel in order to notify the host that such a value is received and used.

On the PC host side, an OLE application is written in Visual Basic to receive and send data. Build the program and run it. Then, run the OLE application embedded in *rtdx.doc* (provided on the accompanying CD-ROM). The CCS should be running when using RTDX. In addition the `readload` and `writeload` channels must be enabled by toggling the checkbox from the menu **Tools → RTDX → Channel Viewer Control**. It can be observed that the CPU load changes as a new `loadVal` is sent from the host OLE program to the DSP.

CHAPTER 11

Lab Project Examples

Four project examples including sinewave generation, IIR filter, filter bank, and pulse amplitude modulation, are presented in this chapter for further exposure to the C6x code writing and optimization. All the files associated with these project examples are placed on the accompanying CD-ROM. The source codes shown in this chapter are for C6713 DSK.

11.1 Sinewave Generation

This project involves generating sinusoidal waveforms based on a difference equation. An oscillatory output, y[n], can be obtained from the following difference equation:

$$y[n] = B_1 * x[n-1] + A_1 * y[n-1] + A_0 * y[n-2] \qquad (11.1)$$

if $B_1 = 1$, $A_0 = -1$, $A_1 = 2\cos(\theta)$ and x[n] is a delta function. Here, this equation is implemented on the DSK board and the generated signal is observed on an oscilloscope. The output frequency is measured and compared with the expected frequency indicated below [1],

$$F = \frac{F_s}{2\pi} \arccos(A_1/2) \qquad (11.2)$$

The coefficient A_1 can be changed to produce different frequencies. Also, the gain B_1 can be altered to see the effect on the generated sinusoid. The interested reader is referred to [1] for theoretical details of sinewave generation.

Project 1: Sinewave Generation

Both the sampling frequency and signal frequency should be defined in order to calculate the angular frequency θ, or the coefficient A_1. Let's consider Fs = 8000 Hz and F = 500 Hz. A delta function input is obtained by initially setting $x[n]$ to 1, and then to 0 after the second iteration. The difference equation is implemented within the ISR, transmit_ISR, where a new sample is generated and exported through the multichannel serial port at every 1/8000 second. The C program for sinewave generation is shown below:

```
#define CHIP_6713 1

#include <stdio.h>
#include <c6x.h>
#include <csl.h>
#include <csl_mcbsp.h>
#include <csl_irq.h>
#include <math.h>

#include "dsk6713.h"
#include "dsk6713_aic23.h"

#define SCALE 1000
#define PI            3.141592

interrupt void transmit_ISR(void);

float y[3], a[2], b1, x;

DSK6713_AIC23_CodecHandle hCodec;
DSK6713_AIC23_Config config = DSK6713_AIC23_DEFAULTCONFIG;
                            // Codec configuration with default settings

main()
{
        float F, Fs, theta;

        DSK6713_init();              // Initialize the board support library
        hCodec = DSK6713_AIC23_openCodec(0, &config);

        MCBSP_FSETS(SPCR1, RINTM, FRM);
        MCBSP_FSETS(SPCR1, XINTM, FRM);
        MCBSP_FSETS(RCR1, RWDLEN1, 32BIT);
        MCBSP_FSETS(XCR1, XWDLEN1, 32BIT);

        DSK6713_AIC23_setFreq(hCodec, DSK6713_AIC23_FREQ_8KHZ);

        // Coefficient Initialization

        F = 500;                          // Signal frequency
        Fs = 8000;                        // Sampling frequency
        theta = (2*PI*F)/Fs;

        a[0] = -1;
        a[1] = 2 * cos(theta);
        b1 = 1;

        // Initial Conditions
```

```
        y[1] = y[2] = 0;
        x = 1;

        IRQ_globalDisable();         // Globally disables interrupts
        IRQ_nmiEnable();             // Enables the NMI interrupt
        IRQ_map(IRQ_EVT_XINT1,15);   // Maps an event to a physical intr number
        IRQ_enable(IRQ_EVT_XINT1);   // Enables the event
        IRQ_globalEnable();          // Globally enables interrupts

        while(1)
        {
        }
}

interrupt void transmit_ISR(void)
{
        int temp;

        y[0] = b1 * x + a[1]*y[1] + a[0]*y[2];

        y[2] = y[1];
        y[1] = y[0];

        x = 0;

        temp = (short) ( y[0] * SCALE );

        MCBSP_write(DSK6713_AIC23_DATAHANDLE, temp << 16 );
}
```

11.2 Cascade IIR Filter

This project example involves designing and implementing IIR filters in their 2nd order cascade form. Let's consider an 8th order Chebyshev lowpass IIR filter in both a direct and a cascade form with the following specifications: a passband from 0 to 1 kHz, a stopband of 1.4–4 kHz, a passband ripple of 1dB, a stopband attenuation of 50 dB, and a sampling frequency of 8 kHz. A C program is first written to implement this filter in both direct and cascade forms. The operation of the program is verified by using a function generator and an oscilloscope. In addition to C, an assembly, a hand-optimized assembly, and a linear assembly code are written to implement the cascade form of the filter. By lowering the Q-format representation of the coefficients, it is shown that the cascade form is less sensitive to quantization as compared with the direct form.

Project 2: Cascade IIR Filter Implementation

Here, we have used the FDATool of MATLAB to obtain the coefficients of the filter. The quantization of the coefficients alters the frequency response, as shown in Figure 11-1. With the second-order cascade filter implementation, the sensitivity to coefficient quantization is significantly reduced. The interested reader is referred to [1] for theoretical details. The frequency response of the second-order cascade implementation is shown in Figure 11-2, and the quantized filter coefficients in Q-15 and single-precision formats are listed in Table 11-1 for each section.

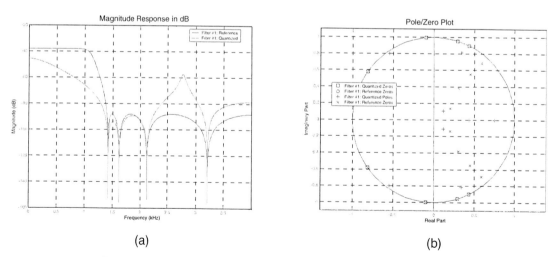

(a)   (b)

**Figure 11-1: Effect of quantization for a single section on
(a) magnitude response (b) pole/zero plot.**

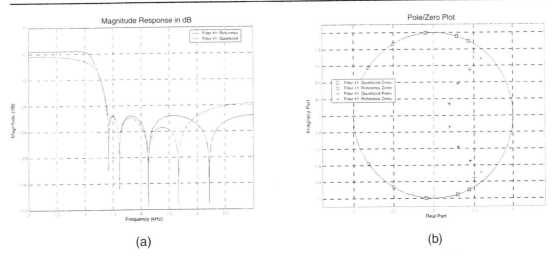

(a)   (b)

Figure 11-2: Effect of quantization for second-order cascade filter on (a) magnitude response (b) pole/zero plot.

Table 11-1: Coefficients of cascaded second-order filters.

Quantized Coefficients in Q-15 format	Designed Coefficients
------- Section 1 -------	
Numerator	
0x7FFF (0.999969482421875)	1.000000000000000000
0x7FFF (0.999969482421875)	1.631899318476804000
0x7FFF (0.999969482421875)	1.000000000000000400
Denominator	
0x7FFF (0.999969482421875)	1.000000000000000000
0xCC01 (-0.406250000000000)	-0.406240975551636410
0x07CA (0.060852050781250)	0.060839816716921374
------- Section 2 -------	
Numerator	
0x7FFF (0.999969482421875)	1.000000000000000000
0x1905 (0.195465087890625)	0.195463825774149170
0x7FFF (0.999969482421875)	1.000000000000001800
Denominator	
0x7FFF (0.999969482421875)	1.000000000000000000
0xB177 (-0.613586425781250)	-0.613596465553117860
0x1E9C (0.239135742187500)	0.239148166557621610
------- Section 3 -------	
Numerator	
0x7FFF (0.999969482421875)	1.000000000000000000
0xB439 (-0.592041015625000)	-0.592045778337563780
0x7FFF (0.999969482421875)	0.999999999999997780
Denominator	
0x7FFF (0.999969482421875)	1.000000000000000000
0x8C7E (-0.902435302734375)	-0.902432041178152230
0x4104 (0.507934570312500)	0.507920458903674770
------- Section 4 -------	
Numerator	
0x7FFF (0.999969482421875)	1.000000000000000000
0x8FC3 (-0.876892089843750)	-0.876905411177924200
0x7FFF (0.999969482421875)	0.999999999999998780
Denominator	
0x7FFF (0.999969482421875)	1.000000000000000000
0x8000 (-1.000000000000000)	-1.180266034077308100
0x68B4 (0.817993164062500)	0.818001830217691680
Scale Value = 0x019A (0.0125407)	

The quantized coefficients are first stated in floating-point format, and then converted into Q-15 format. This conversion needs to be done before enabling the interrupt service routine. In the interrupt service routine, serialPortRcvISR, the gain factor is multiplied with the input to avoid possible overflows followed by four sections of the IIR filter. Each section of the IIR filter simply consists of dot-products and buffering. The output from a previous section becomes the input of a following section. The accuracy of the Q-format representation, Qnum in the code, and the maximum positive number corresponding to the current representation, MAX_POS, for example, 0x7FFF for Q-15 format, should be changed for other Q-format representations.

```
#define CHIP_6713 1

#include <stdio.h>
#include <c6x.h>
#include <csl.h>
#include <csl_mcbsp.h>
#include <csl_irq.h>

#include "dsk6713.h"
#include "dsk6713_aic23.h"

#define N           3           // Number of coefficients
#define SCALE 2                 // Scale factor
#define Q           15
#define MAX_POS     0x7FFF

DSK6713_AIC23_CodecHandle hCodec;
DSK6713_AIC23_Config config = DSK6713_AIC23_DEFAULTCONFIG;
                               // Codec configuration with default settings


// Coefficients of second order IIR filter
float b1_f[N] = {0.999969482421875,  0.999969482421875,  0.999969482421875 };
float a1_f[N] = {0.999969482421875, -0.406250000000000,  0.060852050781250 };
float b2_f[N] = {0.999969482421875,  0.195465087890625,  0.999969482421875 };
float a2_f[N] = {0.999969482421875, -0.613586425781250,  0.239135742187500 };
float b3_f[N] = {0.999969482421875, -0.592041015625000,  0.999969482421875 };
float a3_f[N] = {0.999969482421875, -0.902435302734375,  0.507934570312500 };
float b4_f[N] = {0.999969482421875, -0.876892089843750,  0.999969482421875 };
float a4_f[N] = {0.999969482421875, -1.000000000000000,  0.817993164062500 };
float G_f = 0.0125407;
short b1[N], a1[N], b2[N], a2[N], b3[N], a3[N], b4[N], a4[N], G;
int x[N], y1[N], y2[N], y3[N], y4[N];

void main()
{
        int i;

        DSK6713_init();                 // Initialize the board support library
        hCodec = DSK6713_AIC23_openCodec(0, &config);

        MCBSP_FSETS(SPCR1, RINTM, FRM);
        MCBSP_FSETS(SPCR1, XINTM, FRM);
        MCBSP_FSETS(RCR1, RWDLEN1, 32BIT);
        MCBSP_FSETS(XCR1, XWDLEN1, 32BIT);

        DSK6713_AIC23_setFreq(hCodec, DSK6713_AIC23_FREQ_8KHZ);

        // Initial conditions
        for( i = 0; i < N; i++ )
        {
                x[i] = 0;
                y1[i] = y2[i] = y3[i] = y4[i] = 0;
        }
```

```
        // Convert coefficients to Q format
        G = MAX_POS * G_f;
        for( i = 0; i < N; i++ )
        {
                b1[i]= MAX_POS * b1_f[i];
                a1[i]= MAX_POS * a1_f[i];
                b2[i]= MAX_POS * b2_f[i];
                a2[i]= MAX_POS * a2_f[i];
                b3[i]= MAX_POS * b3_f[i];
                a3[i]= MAX_POS * a3_f[i];
                b4[i]= MAX_POS * b4_f[i];
                a4[i]= MAX_POS * a4_f[i];
        }

        // Interrupt Configuration
        IRQ_globalDisable();                // Globally disables interrupts
        IRQ_nmiEnable();                    // Enables the NMI interrupt
        IRQ_map(IRQ_EVT_RINT1,15);          // Maps an event to a physical intr
        IRQ_enable(IRQ_EVT_RINT1);          // Enables the event
        IRQ_globalEnable();                 // Globally enables interrupts

        while(1)
        {
        }
}

interrupt void serialPortRcvISR(void)
{
        int temp;

        temp = MCBSP_read(DSK6713_AIC23_DATAHANDLE);      // Read sample
        temp = temp >> (31- Q);

        x[2] = x[1];                                 // Circulate input
        x[1] = x[0];
        x[0] = _mpy(temp,G) >> Q;                    // Multiply gain factor

        //      IIR filter : Section 1
        y1[2] = y1[1];// Circulate output(input for following section)
        y1[1] = y1[0];
        y1[0] = (_mpy(b1[0],x[0]) >> Q) + (_mpy(b1[1],x[1]) >> Q)
        + (_mpy(b1[2],x[2]) >> Q) - (_mpy(a1[1],y1[1]) >> Q)
        - (_mpy(a1[2],y1[2]) >> Q);

        //      IIR filter : Section 2
        y2[2] = y2[1];// Circulate output(input for following section)
        y2[1] = y2[0];
        y2[0] = (_mpy(b2[0],y1[0]) >> Q) + (_mpy(b2[1],y1[1]) >> Q)
                + (_mpy(b2[2],y1[2]) >> Q) - (_mpy(a2[1],y2[1]) >> Q)
                - (_mpy(a2[2],y2[2]) >> Q);

        //      IIR filter : Section 3
        y3[2] = y3[1];// Circulate output(input for following section)
        y3[1] = y3[0];
        y3[0] = (_mpy(b3[0],y2[0]) >> Q) + (_mpy(b3[1],y2[1]) >> Q)
```

```
        + (_mpy(b3[2],y2[2]) >> Q) - (_mpy(a3[1],y3[1]) >> Q)
        - (_mpy(a3[2],y3[2]) >> Q);

        //      IIR filter : Section 4
        y4[2] = y4[1];// Circulate output
        y4[1] = y4[0];
        y4[0] = (_mpy(b4[0],y3[0]) >> Q) + (_mpy(b4[1],y3[1]) >> Q)
                + (_mpy(b4[2],y3[2]) >> Q) - (_mpy(a4[1],y4[1]) >> Q)
                - (_mpy(a4[2],y4[2]) >> Q);

        temp = y4[0] * SCALE;                   // Amplify the output
        temp = temp << (31-Q);

        MCBSP_write(DSK6713_AIC23_DATAHANDLE, temp);
}
```

The code is rewritten in assembly. Each IIR filter section is replaced with an assembly function, which is called by the following code line:

```
    y1[0] = iir(x, b1, y1, a1);    // Call IIR filter written in assembly
```

The above function call loads the current input, previous outputs, and filter coefficients, and calculates the dot-products corresponding to the numerator and denominator. The numerator sum is subtracted from the denominator sum for the final output. The assembly code for doing so is given below:

```
        .global      _iir          ; Simple iir filter implementation
        .sect ".iir"
_iir:
        MVK    .S1    3, A2         ; counter
        ZERO   .S1    A9            ; BSUM

loop1:
        LDW    .D1    *A4++,A5      ; Load input sample
        LDH    .D2    *B4++,B5      ; Load b coefficient
        NOP    4
        MPY    .M1x   A5, B5, A8    ; b * input
        NOP
```

```
          SHR    .S1    A8, 15, A8
          ADD    .S1    A8, A9, A9

   [A2] SUB    .S1    A2, 1, A2
   [A2] B      .S2    loop1
        NOP    5

          MVK    .S1    2, A2          ; counter
          ZERO   .S2    B9             ; ASUM

loop2:
          LDW    .D1    *++A6,A7       ; Load output sample
          LDH    .D2    *++B6,B7       ; Load a coefficient
          NOP    4
          MPY    .M2x   A7, B7, B8     ; a * output
          NOP
          SHR    .S2    B8, 15, B8
          ADD    .S2    B8, B9, B9

   [A2] SUB    .S1    A2, 1, A2
   [A2] B      .S2    loop2
        NOP    5

          SUB    .L1    A9, B9, A4     ; BSUM - ASUM

          B      .S2    B3
          NOP    5
```

The number of cycles corresponding to one section of the IIR filter is shown in Table 11-2 for different builds. As can be seen from this table, the C implementation provides a faster outcome as compared with the assembly version, and the linear assembly version provides a slower outcome as compared with the assembly version. The main reason for this is that there are too few repetitions, 3 at most, of the loop. The C code is already fully optimized by using assembly intrinsics and not using any loop structure, while the linear assembly and the assembly are implemented with loop structures. The difference between optimized C and hand-optimized assembly is due to the fact that calling assembly function in C incorporates several procedures such as passing arguments from C to assembly, storing return address, branching to assembly routine, returning output value, and branching back from assembly routine. As the number of repetition increases, the efficiency of the linear assembly and hand-optimized assembly as compared with the C version becomes more noticeable.

Table 11-2: Number of cycles for different builds.

Build type	Number of cycles
C without optimization (none/-o3)	38 / 27
Linear assembly (none/-o3)	152 / 183
Simple assembly	132
Hand-optimized assembly	51

11.3 Filter Bank

This project involves the implementation of a 2-channel filter bank system, as shown in Figure 11-3. Such a system is used to lower the bit rate for speech coding applications. Consider an 8 kHz sampling rate for the input signal, i.e. the input signal having a 4 kHz bandwidth (speech signal). The interested reader is referred to [1] for theoretical details of filter banks.

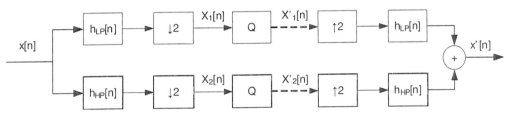

Figure 11-3: 2-channel Filter bank.

By reducing the number of quantization bits in each channel, one can hear the effect on the reconstructed signal. Recorded speech signals can be used for testing. The filter bank implementation is done by writing a C, an optimized assembly, and a linear assembly code. The number of cycles and code size associated with each coding approach are then compared.

Project 3: Filter Bank Implementation

We have considered a 32-coefficient lowpass filter here. If the size of the input file is too large for the available internal memory, it should be loaded to the external memory. The following line allocates space for the specified variable in the user designated section "my_sect", which is defined as part of the external memory in the linker command file,

```
#pragma DATA_SECTION(f_input, "my_sect")
```

The data file is then loaded by choosing **File** → **Data** → **Load**, selecting **Float(*.dat)** in the **Files of type** field, and then selecting *signal.dat*. By clicking open, the **Loading File into Memory** window is brought up. Type 0x80000000 in the **Address** and 0x00000100 in the **Length** field, respectively.

The code shown below includes the analysis filter, down-sampler, quantizer, up-sampler, and synthesis filter. The coefficients of the highpass filter are obtained simply by changing the sign of the odd coefficients of the lowpass filter.

```
#include <stdio.h>

#define N 32
#define SIG_LEN    0x0100        // 256
#define Qbit       0

float f_coef[N]={ -0.017824925492008,  -0.035774695205120,  0.022767212078393,
  0.009949352454172, -0.014276325762842, -0.018785999870724, 0.020967756230837,
  0.023573942319073, -0.027510480213313, -0.032452910146451, 0.039131601497448,
  0.048479743981073, -0.063129344823646, -0.089156565459054, 0.149537214887298,
  0.449964294508326,  0.449964294508326,  0.149537214887298, -0.089156565459054,
 -0.063129344823646,  0.048479743981073,  0.039131601497448, -0.032452910146451,
 -0.027510480213313,  0.023573942319073,  0.020967756230837, -0.018785999870724,
 -0.014276325762842,  0.009949352454172,  0.022767212078393, -0.035774695205120,
 -0.017824925492008};

#pragma DATA_SECTION(f_input, "my_sect")
#pragma DATA_SECTION(input, "my_sect")
#pragma DATA_SECTION(L_intermediate, "my_sect")
#pragma DATA_SECTION(H_intermediate, "my_sect")
#pragma DATA_SECTION(output, "my_sect")
float f_input[SIG_LEN];
int   input[SIG_LEN], output[SIG_LEN],
      L_intermediate[SIG_LEN], H_intermediate[SIG_LEN];

void main()
{
      int   L_temp, H_temp;
      short i, k;
      short L_coef[N], H_coef[N];
      short samples[N];
      short L_samples[N], H_samples[N];

      // Convert input data into Q-15 format
      for( i = 0 ; i < SIG_LEN; i++ )
            input[i] = 0x7FFF * f_input[i];

      // Convert coefficients of filter into Q-15 format
      for( i = 0 ; i < N; i++ )
      {
            L_coef[i] = 0x7FFF * f_coef[i];             // Lowpass filter
            H_coef[i] = i%2 ? -L_coef[i] : L_coef[i];   // Highpass filter
      }
```

```
// Initial condition
for( i = 0 ; i < N; i++ )
      samples[i] = 0;

// Analysis filters
for( k = 0; k < SIG_LEN ; k++ )
{
      for(i = N-1 ; i > 0 ; i-- )
      {
            samples[i] = samples[i-1];
      }
      samples[0] = input[k];

      L_temp = H_temp = 0;
      for( i = 0 ; i < N ; i++ )
      {
            L_temp += ( _mpy(samples[i], L_coef[i]) >> 15 );
            H_temp += ( _mpy(samples[i], H_coef[i]) >> 15 );
      }

      // Quantizer
      L_intermediate[k] = L_temp >> Qbit;
      H_intermediate[k] = H_temp >> Qbit;
}

// Decimation & Interpolation
for( k = 1; k < SIG_LEN ; k = k + 2 )
{
      L_intermediate[k] = 0;
      H_intermediate[k] = 0;
}

// --------------------------------------------------------------------

// Initial condition
for( i = 0 ; i < N ; i++ )
{
      L_samples[i] = 0;
      H_samples[i] = 0;
}

// Synthesis filters
for( k = 0; k < SIG_LEN ; k++ )
{
      for( i = N-1 ; i > 0; i-- )               // Circulate buffer
      {
            L_samples[i] = L_samples[i-1];
            H_samples[i] = H_samples[i-1];
      }
      L_samples[0] = L_intermediate[k];
      H_samples[0] = H_intermediate[k];

      L_temp = H_temp = 0;
      for( i = 0 ; i < N ; i++ )                // FIR filter
      {
            L_temp += ( _mpy( L_samples[i], L_coef[i] ) >> 15 );
            H_temp += ( _mpy( H_samples[i], H_coef[i] ) >> 15 );
```

```
        }

            output[k] = L_temp + H_temp;   // Add outputs from both filter
    }
}
```

Both the lowpass and highpass filters are implemented as FIR filters. One can study the effect of quantization by changing the value of Qbit in the code. The code size and number of cycles for different builds are shown in Table 11-3. From this table, it can be seen that there is a trade-off between the code size and number of cycles.

Table 11-3: Code size and number of cycles of analysis filter for different builds.

Build type	Code size	Number of cycles
C without optimization (none/-o3)	136 / 336	1284 / 87
Linear assembly (none/-o3)	60 / 296	731 / 124
Simple assembly	64	608
Hand-optimized assembly	256	95

11.4 Pulse Amplitude Modulation (PAM)

In this project, a C program is written to generate pseudo-random sequences (PN) using a linear feedback shift register sequence generator, shown in Figure 11-4. The connection polynomial is given by

$$h(D) = 1 + D^2 + D^5 \tag{11.3}$$

where the summation represents modulo 2 additions. For pulse shaping, a raised-cosine FIR filter can be designed by using the MATLAB function 'rcosfir' with the parameters roll-off factor set to 0.125, extent of the filter to 4, and oversampling rate to 4. The PN sequence is taken as the input of the raised cosine filter, and the polyphase filter bank approach is used for code efficiency. Refer to [2] for theoretical details.

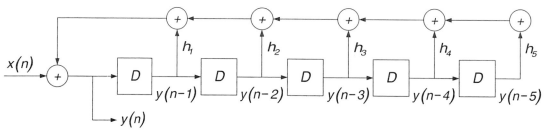

Figure 11-4: Pseudo Noise generation.

Lab 12: PAM Implementation

The pulse amplitude modulation is achieved by a PN sequence followed by the raised cosine filter for pulse shaping. The PN sequence is implemented by the linear feedback shift registers, and the raised cosine filter is implemented by using appropriate coefficients. For demonstration purposes, a 5 stage shift register structure generating random sequences with period $31 (= 2^5 - 1)$ is chosen. The array g [ ] contains the coefficients of the raised cosine filter designed by using MATLAB. All the polyphase filters take the same PN sample as their input, and then calculate dot-products based on their own coefficients. In the code, the `switch-case` statement is used for the polyphase implementation. The code is shown below:

```
#define CHIP_6713 1

#include <stdio.h>
#include <math.h>

#include <c6x.h>
#include <csl.h>
#include <csl_mcbsp.h>
#include <csl_irq.h>

#include "dsk6713.h"
#include "dsk6713_aic23.h"

#defineAMP          0x2fff

#defineRATE    4
#define N_T    4
#define BUF_LEN      33 // (2*N_T*RATE+1)

#define N 5    // length of shift register
short y[N+1] = { 1, 0, 0, 0, 0, 0 };

short a[9];
float g[BUF_LEN]={ -0.0000, -0.0234, -0.0407, -0.0351, 0.0000, 0.0511, 0.0866,
0.0735, -0.0000, -0.1071, -0.1856, -0.1642, 0.0000, 0.2904, 0.6274, 0.8970,
1.0000, 0.8970, 0.6274, 0.2904, 0.0000, -0.1642, -0.1856, -0.1071, -0.0000,
0.0735, 0.0866, 0.0511, 0.0000, -0.0351, -0.0407, -0.0234, -0.0000 };

static short j = RATE;

DSK6713_AIC23_CodecHandle hCodec;
DSK6713_AIC23_Config config = DSK6713_AIC23_DEFAULTCONFIG;

interrupt void serialPortRcvISR(void);

void main()
{
      int i;

      DSK6713_init();              // Initialize the board support library
      hCodec = DSK6713_AIC23_openCodec(0, &config);
```

```
        MCBSP_FSETS(SPCR1, RINTM, FRM);
        MCBSP_FSETS(SPCR1, XINTM, FRM);
        MCBSP_FSETS(RCR1, RWDLEN1, 32BIT);
        MCBSP_FSETS(XCR1, XWDLEN1, 32BIT);

        DSK6713_AIC23_setFreq(hCodec, DSK6713_AIC23_FREQ_8KHZ);

        // Ininitialize PN sequence;
        for ( i = 0; i < 9; i++ )
        {        a[i] = 0;        }

        IRQ_globalDisable();            // Globally disables interrupts
        IRQ_nmiEnable();                // Enables the NMI interrupt
        IRQ_map(IRQ_EVT_XINT1,15);      // Maps an event to a physical interrupt
        IRQ_enable(IRQ_EVT_XINT1);      // Enables the event
        IRQ_globalEnable();             // Globally enables interrupts

        while(1)
        {
        }
}

interrupt void serialPortRcvISR()
{
        int temp;
        float ftemp;
        short i, a0;

        // PN-sequence generation

        if ( j >= RATE )       // If one period is over, generate next PN value.
        {
                j = 0;

                for( i = N ; i > 0 ; i--)
                {
                        y[i] = y[i-1];
                }

                y[0] = 0 ^ y[2] ^ y[5];      // h(D)= 1 + D^2 + D^5,
                                             // input is assumed as 0.
                a0 = 2 * y[0] - 1;

                for ( i = 8 ; i > 0; i-- )
                {
                        a[i] = a[i-1];
                }
                a[0] = a0;
        }

        // Raised cosine FIR filter
```

```
switch (j)
{
case 0 : { ftemp = a[0]*g[0] + a[1]*g[4] + a[2]*g[8] + a[3]*g[12]
          + a[4]*g[16] + a[5]*g[20] + a[6]*g[24] + a[7]*g[28] + a[8]*g[32];
          break; }
case 1 : { ftemp = a[0]*g[1] + a[1]*g[5] + a[2]*g[9] + a[3]*g[13]
          + a[4]*g[17] + a[5]*g[21] + a[6]*g[25] + a[7]*g[29]; break; }
case 2 : { ftemp = a[0]*g[2] + a[1]*g[6] + a[2]*g[10] + a[3]*g[14]
          + a[4]*g[18] + a[5]*g[22] + a[6]*g[26] + a[7]*g[30]; break; }
case 3 : { ftemp = a[0]*g[3] + a[1]*g[7] + a[2]*g[11] + a[3]*g[15]
          + a[4]*g[19] + a[5]*g[23] + a[6]*g[27] + a[7]*g[31]; break; }
}

temp = (int) (AMP * ftemp); // Scale

j++;

MCBSP_write(DSK6713_AIC23_DATAHANDLE, temp << 16 );
}
```

The effect of roll-off factor, which determines the bandwidth of the pulse shape, can be verified by deploying different coefficient sets corresponding to various roll-off factors between 0 and 1. The DSK output can be connected to an oscilloscope to see the eye diagram on the oscilloscope.

Bibliography

[1] J. Proakis and D. Manolakis, *Digital Signal Processing: Principles, Algorithms, and Applications*, Prentice-Hall, 1996.

[2] S. Tretter, *Communication System Design Using DSP Algorithms: With Laboratory Experiments for the TMS320C6701 and TMS320C6711*, Kluwer Academic Publishers, 2003.

Appendix A:
Quick Reference Guide

A.1 List of C6x Instructions

.L Unit

Instruction	Description
ABS	Integer absolute value with saturation
ADD(U)	Signed(unsigned) addition without saturation
AND	Bitwise AND
CMPEQ	Integer compare for equality
CMPGT	Signed integer compare for greater than
CMPGTU	Unsigned integer compare for greater than
CMPLT	Signed integer compare for less than
CMPLTU	Unsigned integer compare for less than
LMBD	Leftmost bit detection
MV	Move from register to register
NEG	Negate
NORM	Normalize integer
NOT	Bitwise NOT
OR	Bitwise OR
SADD	Integer addition with saturation to result size
SAT	Saturate a 40-bit Integer to a 32-bit Integer
SSUB	Integer subtraction with saturation to result size
SUB(U)	Signed (unsigned) integer subtraction without saturation
SUBC	Conditional integer subtract and shift – used for division
XOR	Exclusive OR
ZERO	Zero a register

.M Unit

MPY	Signed integer multiply 16lsb × 16 lsb
MPYU	Unsigned integer multiply 16lsb × 16lsb
MPYUS	Integer multiply (unsigned) 16lsb × (signed) 16lsb
MPYSU	Integer multiply (signed) 16lsb × (unsigned) 16lsb
MPYH	Signed integer multiply 16msb × 16msb
MPYHU	Unsigned integer multiply 16msb × 16msb
MPYHUS	Integer multiply (unsigned) 16msb × (signed) 16msb
MPYHSU	Integer multiply (signed) 16msb × (unsigned) 16msb
MPYHL	Signed multiply high low 16msb × 16lsb
MPYHLU	Unsigned multiply high low 16msb × 16lsb
MPYHULS	Multiply high unsigned low signed (unsigned) 16msb × (signed) 16lsb
MPYHSLU	Multiply high signed low unsigned (signed) 16msb × (unsigned) 16lsb
MPYLH	Signed multiply low high 16lsb × 16msb
MPYLHU	Unsigned multiply low high 16lsb × 16msb
MPYLUHS	Multiply low unsigned high signed (unsigned) 16lsb × (signed) 16msb
MPYLSHU	Multiply low signed high unsigned (signed) 16lsb × (unsigned) 16msb

SMPY	Integer multiply with left shift and saturation
SMPYHL	Integer multiply high low with left shift and saturation
SMPYLH	Integer multiply low high with left shift and saturation
SMPYH	Integer multiply high with left shift and saturation

.S Unit	
ADD	Signed integer addition without saturation
ADDK	Integer addition using signed 16-bit constant
ADD2	Two 16-bit integer adds on upper and lower register halves
AND	Bitwise AND
B disp	Branch using a displacement
B IRP	Branch using an Interrupt return pointer
B NRP	Branch using a NMI return pointer
B reg	Branch using a register
CLR	Clear a bit field
EXT(U)	Extract and sign-extend(zero-extend) a bit field
MV	Move from register to register
MVC	Move between the control file and register file
MVK	Move a 16-bit signed constant into a register and sign extend
MVKH	Move 16-bit constant into the upper bits of a register
MVKLH	Move 16-bit constant into the upper bits of a register
NEG	Negate
NOT	Bitwise NOT
OR	Bitwise OR
SET	Set a bit field
SHL	Arithmetic shift left
SHR	Arithmetic shift right
SHRU	Logical shift right
SHRL	Logical shift left
SUB(U)	Signed (Unsigned) integer subtraction without saturation
SUB2	Two 16-bit Integer subtracts on upper and lower register halves
XOR	Exclusive OR
ZERO	Zero a register

.D Unit	
ADD	Signed integer addition without saturation
ADDAB/ADDAH/ADDAW	Integer addition using addressing mode
LDB(U)/LDH(U)/ LDW	Load from memory with a 5-bit unsigned constant offset or register offset
LDB(U)/LDH(U)/ LDW (15-bit offset)	Load from memory with a 15-bit constant offset
MV	Move from register to register
STB/STH/STW	Store to memory with a register offset or 5-bit unsigned constant offset
STB/STH/STW (15-bit offset)	Store to memory with a 15-bit offset
SUB	Signed integer subtraction without saturation
SUBAB/SUBAH/ SUBAW	Integer subtraction using addressing mode
ZERO	Zero a register

A.2 List of C67x Floating-Point Instructions

.L Unit

Instruction	Description
ADDDP	Double-precision floating-point addition
ADDSP	Single-precision floating-point absolute value
DPINT	Convert double-precision floating-point value to integer
DPSP	Convert double-precision floating-point value to single-precision floating-point value
DPTRUNC	Convert double-precision floating-point value to integer with truncation
INTDP	Convert integer to double-precision floating-point value
INTDPU	Convert integer to double-precision floating-point value (unsigned)
INTSP	Convert integer to single-precision floating-point value
INTSPU	Convert integer to single-precision floating-point value (unsigned)
SPINT	Convert single-precision floating-point value to integer
SPTRUNC	Convert single-precision floating-point value to integer with truncation
SUBSP	Single-precision floating-point subtract
SUBDP	Double-precision floating-point subtract

.M Unit

MPYSP	Single-precision floating-point multiply
MPYDP	Double-precision floating-point multiply
MPYI	32-bit integer multiply - result in lower 32 bits
MPYID	32-bit integer multiply - 64-bit result

.S Unit

ABSSP	Single-precision floating-point absolute value
ABSDP	Double-precision floating-point absolute value
CMPGTSP	Single-precision floating-point compare for greater than
CMPEQSP	Single-precision floating-point compare for equality
CMPLTSP	Single-precision floating-point compare for less than
CMPGTDP	Double-precision floating-point compare for greater than
CMPEQDP	Double-precision floating-point compare for equality
CMPLTDP	Double-precision floating-point compare for less than
RCPSP	Single-precision floating-point reciprocal approximation
RCPDP	Double-precision floating-point reciprocal approximation
RSQRSP	Single-precision floating-point square-root reciprocal approximation
RSQRDP	Double-precision floating-point square-root reciprocal approximation
SPDP	Convert Single precision floating-point value to double-precision floating-point value

.D Unit

ADDAD	Integer addition using doubleword addressing mode
LDDW	Load doubleword from memory with an offset

A.3 Registers and Memory Mapped Registers†

Addressing Mode Register (AMR)

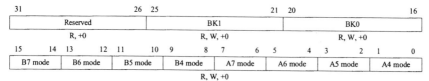

Control Status Register (CSR)

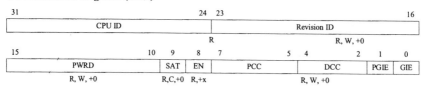

Interrupt Flag Register (IFR)

15	14	13	12	11	10	9	8	7	6	5	4	3	2	1	0
IF15	IF14	IF13	IF12	IF11	IF10	IF9	IF8	IF7	IF6	IF5	IF4	rsv	rsv	NMIF	0

R, +0

Interrupt Set Register (ISR)

15	14	13	12	11	10	9	8	7	6	5	4	3	2	1	0
IS15	IS14	IS13	IS12	IS11	IS10	IS9	IS8	IS7	IS6	IS5	IS4	rsv	rsv	rsv	rsv

W

Interrupt Clear Register (ICR)

15	14	13	12	11	10	9	8	7	6	5	4	3	2	1	0
IC15	IC14	IC13	IC12	IC11	IC10	IC9	IC8	IC7	IC6	IC5	IC4	rsv	rsv	rsv	rsv

W

Interrupt Enable Register (IER)

15	14	13	12	11	10	9	8	7	6	5	4	3	2	1	0
IE15	IE14	IE13	IE12	IE11	IE10	IE9	IE8	IE7	IE6	IE5	IE4	rsv	rsv	NMIE	1

R, W, +0   R, 1

Interrupt Service Table Pointer (ISTP)

NMI Return Pointer (NRP)

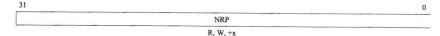

Interrupt Return Pointer (IRP)

Note: Bits not shown are **reserved**.

C6x11 DSK MMR

EMIF

Global Control	1800000
CE0 Space Control	1800008
CE1 Space Control	1800004
CE2 Space Control	1800010
CE3 Space Control	1800014
SDRAM Control	1800018
SDRAM Refresh Period	180001C
SDRAM extension	1800020

Cache

Cache Configuration	1840000
L2 Flush Base Address	1844000
L2 Flush Word Count	1844004
L2 Clean Base Address	1844010
L2 Clean Word Count	1844014
L1P Flush Base Address	1844020
L1P Flush Word Count	1844024
L1D Flush Base Address	1844030
L1D Flush Word Count	1844034
L2 Flush	1845000
L2 Clean	1845004
Memory Attribute	1848200-18482CC

EDMA

Options	1A00000
Source Address	1A00004
Count	1A00008
Destination Address	1A0000C
Index	1A00010
Count Reload./Link Address	1A00014
Priority Queue Status	1A0FFE0
Channel Interrupt Pending	1A0FFE4
Channel Interrupt Enable	1A0FFE8
Channel Chain Enable	1A0FFEC
Event	1A0FFF0
Event Enable	1A0FFF4
Event Clear	1A0FFF8
Event Set	1A0FFFC

QDMA

Options	2000000
Source Address	2000004
Transfer Count	2000008
Destination Address	200000C
Index	2000010
Options Pseudo	2000020
Source Address Pseudo	2000024
Transfer Count Pseudo	2000028
Destination Address Pseudo	200002C

HPI

Control Register	1880000

Interrupts

Multiplexer High	19C0000
Multiplexer Low	19C0004
External Interrupt Polarity	19C0008

Index Pseudo	2000030	

McBSP	0	1
DRR	18C0000	1900000
DXR	18C0004	1900004
Control Register	18C0008	1900008
Receive Control Register	18C000C	190000C
Transmit Control Register	18C0010	1900010
Sample Rate Generator Register	18C0014	1900014
Multichannel Register	18C0018	1900018
Receive Channel Enable Register	18C001C	190001C
Transmit Channel Enable Register	18C0020	1900020
Pin Control Register	18C0024	1900024

Timers	0	1
Control	1940000	1980000
Period	1940004	1980004
Counter	1940008	1980008

C64x DSK MMR

EMIF	A	B
Global Control	1800000	1A80000
CE1 Space Control	1800008	1A80008
CE0 Space Control	1800004	1A80004
CE2 Space Control	1800010	1A80010
CE3 Space Control	1800014	1A80014
SDRAM Control	1800018	1A80018
SDRAM Refresh Period	180001C	1A8001C
SDRAM extension	1800020	1A80020
CE1 Space Secondary Control	1800044	1A80044
CE0 Space Secondary Control	1800048	1A80048
CE2 Space Secondary Control	1800050	1A80050
CE3 Space Secondary Control	1800054	1A80054

HPI	
Control Register	1880000
Address Write	1880004
Address Read	1880008

Interrupts	
Multiplexer High	19C0000
Multiplexer Low	19C0004
External Interrupt Polarity	19C0008

Cache	
Cache Configuration	1840000
L2 Allocation 0	1842000
L2 Allocation 1	1842004
L2 Allocation 2	1842008
L2 Allocation 3	184200C
L2 Flush Base Address	1844000
L2 Flush Word Count	1844004
L2 Clean Base Address	1844010
L2 Clean Word Count	1844014
L1P Flush Base Address	1844020
L1P Flush Word Count	1844024
L1D Flush Base Address	1844030
L1D Flush Word Count	1844034
L2 Flush	1845000
L2 Clean	1845004
Memory Attribute	1848180-18481BC
	1848200-18482FC

EDMA	
Options	1A00000
Source Address	1A00004
Count	1A00008
Destination Address	1A0000C
Index	1A00010
Count Reload./Link Address	1A00014
Priority Queue Status	1A0FFE0
Priority Queue Allocation 0	1A0FFC0
Priority Queue Allocation 1	1A0FFC4
Priority Queue Allocation 2	1A0FFC8
Priority Queue Allocation 3	1A0FFCC
Channel Interrupt Pending Low	1A0FFE4
Channel Interrupt Pending High	1A0FFA4
Channel Interrupt Enable Low	1A0FFE8
Channel Interrupt Enable High	1A0FFA8

Channel Chain Enable Low	1A0FFEC
Channel Chain Enable High	1A0FFAC
Event Low	1A0FFF0
Event High	1A0FFB0
Event Enable Low	1A0FFF4
Event Enable High	1A0FFB4
Event Polarity Low	1A0FFDC
Event Polarity High	1A0FF9C
Event Clear Low	1A0FFF8
Event Clear High	1A0FFB8
Event Set Low	1A0FFFC
Event Set High	1A0FFBC

QDMA

Options	2000000
Source Address	2000004
Transfer Count	2000008
Destination Address	200000C
Index	2000010
Options Pseudo	2000020
Source Address Pseudo	2000024
Transfer Count Pseudo	2000028
Destination Address Pseudo	200002C
Index Pseudo	2000030

McBSP	**0**	**1**	**2**
DRR	18C0000	1900000	1A40000
DXR	18C0004	1900004	1A40004
Control Register	18C0008	1900008	1A40008
Receive Control Register	18C000C	190000C	1A4000C
Transmit Control Register	18C0010	1900010	1A40010
Sample Rate Generator Register	18C0014	1900014	1A40014
Multichannel Register	18C0018	1900018	1A40018
Receive Channel Enable Register	18C001C	190001C	1A4001C
Transmit Channel Enable Register	18C0020	1900020	1A40020
Pin Control Register	18C0024	1900024	1A40024

Timers	**0**	**1**	**2**
Control	1940000	1980000	1AC0000
Period	1940004	1980004	1AC0004
Counter	1940008	1980008	1AC0008

CPLD

User	60000000
Daughter Card	60000004
Version	60000010
Miscellaneous	60000018

C6x01 EVM MMR

EMIF

Global Control	1800000
CE0 Space Control	1800008
CE1 Space Control	1800004
CE2 Space Control	1800010
CE3 Space Control	1800014
SDRAM Control	1800018
SDRAM Refresh Period	180001C
SDRAM extension	1800020

HPI

Control Register	1880000

Interrupts

Multiplexer High	19C0000
Multiplexer Low	19C0004
External Interrupt Polarity	19C0008

DMA

	Ch. 0	Ch. 1	Ch. 2	Ch. 3
Primary Control	1840000	1840040	1840004	1840044
Secondary Control	1840008	1840048	184000C	184004C
Source Address	1840010	1840050	1840014	1840054
Destination Address	1840018	1840058	184001C	184005C
Transfer Counter	1840020	1840060	1840024	1840064
Global Reload A	1840028			
Global Reload B	184002C			
Global Index A	1840030			
Global Index B	184003C			
Global Index C	1840068			
Global Index D	184006C			
Auxiliary Control	1840070			

McBSP

	0	1
DRR	18C0000	1900000
DXR	18C0004	1900004
Control Register	18C0008	1900008
Receive Control Register	18C000C	190000C
Transmit Control Register	18C0010	1900010
Sample Rate Generator Register	18C0014	1900014
Multichannel Register	18C0018	1900018
Receive Channel Enable Register	18C001C	190001C
Transmit Channel Enable Register	18C0020	1900020
Pin Control Register	18C0024	1900024

Timers

	0	1
Control	1940000	1980000
Period	1940004	1980004
Counter	1940008	1980008

C6x_MMR.ASM[†]

Memory Mapped Registers

This file must be included in each .asm referencing
mmr register names. It can be included by adding
this line to the top of .asm file:

```
.include c6x_mmr.asm
```

Using the names below simplifies access to peripheral mmr registers.
Here is an example to write all F's into the CE1 and CE2 EMIF
space control registers:

```
MVK     .S1    0FFFFh, A0
MVKLH   .S1    0FFFFh, A0
MVK     .S1    EMIF, A1
MVKH    .S1    EMIF, A1
STW     .D1    A0, *+A1[CE1]
STW     .D1    A0, *+A1[CE2]
```

;Peripheral	Addr/Offset		Register
EMIF	.equ	01800000h	; EMIF global control
CE1	.equ	1	; EMIF CE1 space control
CE0	.equ	2	; EMIF CE0 space control
CE2	.equ	4	; EMIF CE2 space control
CE3	.equ	5	; EMIF CE3 space control
SDRAM	.equ	6	; EMIF SDRAM control
REFRESH	.equ	7	; EMIF SDRAM refresh period
DMA	.equ	01840000h	; Top of DMA registers
DMA0pc	.equ	0	; DMA0 primary control
DMA2pc	.equ	1	; DMA2 primary control
DMA0sc	.equ	2	; DMA0 secondary control
DMA2sc	.equ	3	; DMA2 secondary control
DMA0src	.equ	4	; DMA0 source address
DMA2src	.equ	5	; DMA2 source address
DMA0dst	.equ	6	; DMA0 destination address
DMA2dst	.equ	7	; DMA2 destination address
DMA0tc	.equ	8	; DMA0 transfer counter
DMA2tc	.equ	9	; DMA2 transfer counter
DMAcountA	.equ	10	; DMA global count reload register A
DMAcountB	.equ	11	; DMA global count reload register B
DMAindexA	.equ	12	; DMA global index register A
DMAindexB	.equ	13	; DMA global index register B
DMAaddrA	.equ	14	; DMA global address register A
DMAaddrB	.equ	15	; DMA global address register B
DMA1pc	.equ	16	; DMA1 primary control
DMA3pc	.equ	17	; DMA3 primary control
DMA1sc	.equ	18	; DMA1 secondary control
DMA3sc	.equ	19	; DMA3 secondary control
DMA1src	.equ	20	; DMA1 source address
DMA3src	.equ	21	; DMA3 source address

DMA1dst	.equ	22	; DMA1 destination address
DMA3dst	.equ	23	; DMA3 destination address
DMA1tc	.equ	24	; DMA1 transfer counter
DMA3tc	.equ	25	; DMA3 transfer counter
DMAaddrC	.equ	26	; DMA global address register C
DMAaddrD	.equ	27	; DMA global address register D
DMAaux	.equ	28	; DMA auxiliary control register
HPI	.equ	01880000h	; HPI control register
McBSP0	.equ	018C0000h	; McBSP0 DRR
McBSP1	.equ	01900000h	; McBSP1 DRR
DRR	.equ	0	; McBSP DRR
DXR	.equ	1	; McBSP DXR
SPCR	.equ	2	; McBSP control register
RCR	.equ	3	; McBSP receive control register
XCR	.equ	4	; McBSP transmit control register
SRGR	.equ	5	; McBSP sample rate generator register
MCR	.equ	6	; McBSP multichannel register
RCER	.equ	7	; McBSP receive channel enable register
CER	.equ	8	; McBSP transmit channel enable register
PCR	.equ	9	; McBSP pin control register
Timer0	.equ	01940000h	; Timer 0
Timer 1	.equ	01980000h	; Timer 1
TimCR	.equ	0	: Timer control register
TimTP	.equ	1	; Timer period
TimTC	.equ	2	; Timer counter
Interrupts	.equ	019C0000h	; Interrupts
IMH	.equ	0	; Interrupt multiplexer high
IML	.equ	1	; Interrupt multiplexer low
IP	.equ	2	; External interrupt polarity

A.4 Compiler Intrinsics†

C Compiler Intrinsic	Assembly Instruction	Description	Device
int **_abs**(int src2); int **_labs**(long src2);	ABS	Returns the saturated absolute value of src2	
int **_add2**(int src1, int src2);	ADD2	Adds the upper and lower halves of src1 to the upper and lower halves of src2 and returns the result.	
uint **_clr**(uint src2, uint csta, uint cstb);	CLR	Clears the specified field in src2. The beginning and ending bits of the field to be cleared are specified by csta and cstb respectively.	
unsigned **_clrr**(uint src1, int src2);	CLR	Clears the specified field in src2. The beginning and ending bits of the field to be cleared are specified by the lower 10 bits of the source register.	
int **_dpint**(double);	DPINT	Converts 64-bit double to 32-bit signed integer, using the rounding mode set by the CSR register	'C67x
int **_ext**(uint src2, uint csta, int cstb);	EXT	Extracts the specified field in src2, sign-extended to 32 bits. The extract is performed by a signed shift right; csta and cstb are the shift left and shift right amounts, respectively.	
int **_extr**(int src2, int src1);	EXT	Exctracts the specified field in src2, sign-extended to 32 bits.	
uint **_extu**(uint src2, uint csta, uint cstb);	EXTU	Extracts the specified field in src2, zero-extended to 32 bits.	
uint **_extur**(uint src2, int src1);	EXTU	Extracts the specified field in src2, zero-extended to 32 bits.	
uint **_ftoi**(float);		Reinterprets the bits in the float as an unsigned integer.	'C67x
uint **_hi**(double);		Returns the high 32 bits of a double as an integer.	'C67x
double **_itod**(uint, uint);		Creates a new double register pair from two unsigned integers	'C67x
float **_itof**(uint);		Reinterprets the bits in the unsigned integer as a float.	'C67x
uint **_lmbd**(uint src1, uint src2);	LMBD	Searches for a leftmost 1 or 0 of src2 determined by the LSB of src1. Returns the number of bits up to the bit change.	
uint **_lo**(double);		Returns the low (even) register of a double register pair as an integer	'C67x
int **_mpy**(int src1, int src2); int **_mpyus**(uint src1, int src2); int **_mpysu**(int src1, uint src2); uint **_mpyu**(uint src1, uint src2);	MPY MPYUS MPYSU MPYU	Multiplies the 16 LSBs of src1 by the 16 MSBs of src2 and returns the result. Values can be signed or unsigned.	
int **_mpyhl**(int src1, int src2); int **_mpyhuls**(uint src1, int src2); int **_mpyhslu**(int src1, uint src2); uint **_mpyhlu**(uint src1, uint src2);	MPYHL MPYHULS MPYHSLU MPYHLU	Multiplies the 16 MSBs of src1 by the 16 LSBs of src2 and returns the result. Values can be signed or unsigned,	
int **_mpylh**(int src1, int src2); int **_mpyluhs**(uint src1, int src2);	MPYLH	Multiplies the 16 LSBs of src1 by the	

`int _mpylshu(int src1, uint src2);` `int _mpylhu(uint src1, uint src2);`	**MPYLUHS** **MPYLSHU** **MPYLHU**	16 MSBs of *src2* and returns the result. Values can be signed or unsigned.
`void _nassert(int);`		Generates no code. Tells the optimizer that the expression declared with the assert function is true; this gives a hint to the optimizer as to what optimizations might be valid.
`uint _norm(int src2);` `uint _lnorm(long src2);`	**NORM**	Returns the number of bits up to the first nonredundant sign bit of src2.
`double _rcpdp(double);`	**RCPDP**	Computes the approximate 64-bit double reciprocal.	`C67x
`float _rcpsp(float);`	**RCPSP**	Computes the approximate 32-bit float reciprocal.	`C67x
`double _rsqrdp(float);`	**RSQRDP**	Computes the approximate 64-bit double reciprocal square root.	`C67x
`float _rsqrsp(float);`	**RSQRSP**	Computes the approximate 32-bit float reciprocal square root.	`C67x
`int _sadd(int src1, int src2);` `long _lsadd(int src1, long src2);`	**SADD**	Adds *src1* to *src2* and saturates the results.
`int _sat(long src2);`	**SAT**	Converts a 40-bit value to an 32-bit value and saturates if necessary.
`uint _set(uint src2, uint csta, uint cstb);`	**SET**	Sets the specified field in *src2* to all 1s and returns the *src2* value. The beginning and ending bits of the field to be set are specified by *csta* and *cstb* respectively.
`unsigned _setr(unsigned, int);`	**SET**	Sets the specified field in *src2* to all 1s and returns the *src2* value. The beginning and ending bits of the field to be set are specified by the lower ten bits of the source register.

`int _smpy(int src1, int src2);`
`int _smpyh(int src1, int src2);`
`int _smpyhl(int src1, int src2);`
`int _smpylh(int src1, int src2);`
	SMPY
	SMPYH
	SMPYHL
	SMPYLH
		Multiplies src1 by src2, left shifts the result by one, and returns the result. If the result is 0x8000 0000, saturates the result to 0x7FFF FFFF.

`int _spint(float);`	**SPINT**	Converts 32-bit float to 32-bit signed integer, using the rounding mode set by the CSR register.	`C67x
`uint _sshl(uint src2, uint src1);`	**SSHL**	Shifts *src2* left by the contents of *src1*, saturates the result to 32-bits, and returns the result.
`int _ssub(int src1, uint src2);` `long _lssub(int src1, long src2);`	**SSUB**	Subtracts src2 from src1, saturates the result size, and returns the result.
`uint _subc(uint src1, uint src2);`	**SUBC**	Conditional subtract divide step.
`int _sub2(int src1, int src2);`	**SUB2**	Subtracts the upper and lower halves of *src2* from the upper and lower halves of *src1*, and returns the result. Any borrowing from the lower half subtract does not affect the upper half subtract.

Note: instructions not specified with a device apply to all 'C6x devices.

C64x Specific Intrinsics

C Compiler Intrinsic	Assembly instruction	Description
`int _add4(int src1, int src2);`	**ADD4**	Performs 2s-complement addition to pairs of packed 8-bit numbers
`ushort & _amem2(void *ptr);`	**LDHU** **STHU**	Allows aligned loads of 2 bytes to memory
`uint & _amem4(void *ptr);`	**LDW** **STW**	Allows aligned loads of 4 bytes to memory
`double & _amemd8(void *ptr);`	**LDDW** **STDW** or **LDW/LDW** **STW/STW**	Allows aligned loads of 8 bytes to memory
`const ushort & _amem2_const(const void *ptr);`	**LDHU**	Allows aligned loads of 2 bytes to memory
`const uint & _amem4_const(const void *ptr);`	**LDW**	Allows aligned loads of 4 bytes to memory
`const double & _amemd8_const(const void *ptr);`	**LDDW** or **LDW/LDW**	Allows aligned loads of 8 bytes to memory
`int _avg2(int src1, int src2);`	**AVG2**	Calculates the average for each pair of signed 16-bit values
`uint _avgu4(uint, uint);`	**AVGU4**	Calculates the average for each pair of unsigned 8-bit values
`uint _bitc4(uint src);`	**BITC4**	For each of the 8-bit quantities in src, the number of 1 bits is written to the corresponding position in the return value.
`uint _bitr(uint src);`	**BITR**	Reverses the order of the bits.
`int _cmpeq2(int src1, int src2);`	**CMPEQ2**	Performs equality comparisons on each pair of 16-bit values. Equality results are packed into the two least-significant bits of the return value.
`int _cmpeq4(int src1, int src2);`	**CMPEQ4**	Performs equality comparisons on each pair of 8-bit values. Equality results are packed into the four least-significant bits of the return value.
`int _cmpgt2(int src1, int src2);`	**CMPGT2**	Compares each pair of 8-bit values. Results are packed into the four least-significant bits of the return value.
`uint _cmpgtu4(uint src1, uint src2);`	**CMPGTU4**	Compares each pair of unsigned 8-bit values. Results are packed into the four least-significant bits of the return value.
`uint _deal(uint src);`	**DEAL**	The odd and even bits of src are extracted into two separate 16-bit values.
`int _dotp2(int src1, int src2);` `double _ldotp2(int src1, int src2);`	**DOTP2** **LDOTP2**	The product of the signed lower 16-bit values in src1 and src2 is added to the product of the signed 16-bit values in src1 and src2. Note: The low and high intrinsics are needed to access each half of the 64-bit integer result.
`int _dotpn2(int src1, int src2);`	**DOTPN2**	The product of the signed lower 16-bit values in src1 and src2 is subtracted from the product of the signed upper 16-bit values in src1 and src2.
`int _dotpnrsu2(int src1, uint src2);`	**DOTPNRSU2**	The product of the unsigned lower 16-bit values in src1 and src2 is subtracted from the product of signed 16-bit values in src1 and src2. 2 to the 15th

`int _dotprsu2(int src1, uint src2);`	**DOTPRSU2**	is added and the result is sign shifted right by 16. The product of the signed first pair of 16-bit values is added to the product of the unsigned second pair of 16-bit values. 2 to the 15th is added and the result is sign shifted right by 16.
`int _dotpsu4(int src1, uint src2);` `uint _dotpu4(uint src1, uint src2);`	**DOTPSU4** **DOTPU4**	For each pair of 8-bit values in src1 and src2, the 8-bit value from src1 is multiplied with the 8-bit value from src2. The four products are summed together.
`int _gmpy4(int src1, int src2);`	**GMPY4**	Performs the galois field multiply on four values in src1 with four parallel values in src2. The four products are packed into the return value.

`int _max2(int src1, int src2);`
`uint _maxu4(uint src1, uint src2);`
`int _min2(int src1, int src2);`
`uint _minu4(uint src1, uint src2);`
	MAX2
	MAX4
	MIN2
	MINU4
		Places the larger/smaller of each pair of values in the corresponding position in the return value. Values can be 16-bit signed or 8-bit unsigned.

`ushort & _mem2(void *ptr);`	**LDB/LDB** **STB/STB**	Allows unaligned loads of 2 bytes to memory
`uint & _mem4(void *ptr);`	**LDNW** **STNW**	Allows unaligned loads of 4 bytes to memory
`double & _memd8(void *ptr);`	**LDNDW** **STNDW**	Allows unaligned loads of 8 bytes to memory
`const ushort & _mem2_const(const void *ptr);`	**LDB/LDB**	Allows unaligned loads of 2 bytes to memory
`const uint & _mem4_const(const void *ptr);`	**LDNW**	Allows unaligned loads of 4 bytes to memory
`const double & _memd8_const(const void *ptr);`	**LDNDW**	Allows unaligned loads of 8 bytes to memory
`double _mpy2(int src1, int src2);`	**MPY2**	Returns the products of the lower and higher 16-bit values in src1 and src2
`double _mpyhi(int src1, int src2);` `double _mpyli(int src1, int src2);`	**MPYHI** **MPYLI**	Produces a 16 by 32 multiply. The result is placed into the lower 48 bits of the returned double. Can use the upper or lower 16 bits of src1.
`int _mpyhir(int src1, int src2);` `int _mpylir(int src1, int src2);`	**MPYHIR** **MPYLIR**	Produces a signed 16 by 32 multiply. The result is shifted right by 15 bits. Can use the upper or lower 16 bits of src1.
`double _mpysu4(int src1, uint src2);` `double _mpyu4(uint src1, uint src2);`	**MPYSU4** **MPYU4**	For each 8-bit quantity in src1 and src2, performs an 8-bit by 8-bit multiply. The four 16-bit results are packed into a double. The results can be signed or unsigned.
`int _mvd(int src);`	**MVD**	Moves the data from the src to the return value over four cycles using the multipler pipeline
`uint _pack2(uint src1, uint src2);` `uint _packh2(uint src1, uint src2);`	**PACK2** **PACKH2**	The lower/upper half-words of src1 and src2 are placed in the return value.
`uint _packh4(uint src1, uint src2);` `uint _packl4(uint src1, uint src2);`	**PACKH4** **PACKL4**	Packs alternate bytes into return value. Can pack high or low bytes.
`uint _packhl2(uint src1, uint src2);` `uint _packlh2(uint src1, uint src2);`	**PACKHL2** **PACKLH2**	The upper/lower halfword of src1 is placed in the upper halfword the return value. The lower/upper halfword of src2 is placed in the lower halfword the return value.
`uint _rotl(uint src2, unint src1);`	**ROTL**	Rotates src2 to the left by the amount in src1
`int _sadd2(int src1, int src2);` `int _saddus2(uint src1, int src2);`	**SADD2** **SADDUS2**	Performs saturated addition between pairs of 16-bit values in src1 and src2. Src1 values can be signed or unsigned.
`uint _saddu4(uint src1, uint src2);`	**SADDU4**	Performs saturated addition between pairs of 8-bit unsigned values in src1 and src2.

`uint _shfl(uint src);`	**SHFL**	The lower 16 bits of src are placed in the even bit positions, and the upper 16 bits of src are placed in the odd bit positions.
`Uint _shlmb(uint src1, uint src2);` `uint _shrmb(uint src1, uint src2);`	**SHLMB** **SHRMB**	Shifts src2 left/right by one byte, and the most/least significant byte of src1 is merged into the least/most significant byte position.
`int _shr2(int src1, uint src2);` `uint _shru2(uint src1, uint src2);`	**SHR2** **SHRU2**	For each 16-bit quantity in src2, the quantity is arithmetically or logically shifted right by src1 number of bits. src2 can contain signed or unsigned values.
`double _smpy2(int src1, int src2);`	**SMPY2**	Performs 16-bit multiplication between pairs of signed packed 16-bit values, with an additional 1 bit left-shift and saturate into a double result.
`int _spack2(int src1, int src2);`	**SPACK2**	Two signed 32-bit values are saturated to 16-bit values and packed into the return value
`uint _spacku4(int src1, int src2);`	**SPACKU4**	Four signed 16-bit values are saturated to 8-bit values and packed into the return value.
`int _sshvl(int src2, int src1);` `int _sshvr(int src2, int src1);`	**SSHVL** **SSHVR**	Shifts src2 to the left/right src1 bits. Saturates the result if the shifted value is greater than MAX_INT or less than MIN_INT.
`int _sub4(int src1, int src2);`	**SUB4**	Performs 2s-complement subtraction between pairs of packed 8-bit values
`int _subabs4(int src1, int src2);`	**SUBABS4**	Calculates the absolute value of the differences for each pair of packed 8-bit values.
`uint _swap4(uint src);`	**SWAP4**	Exchanges pairs of bytes (an endian swap) within each 16-bit value
`uint _unpkhu4(uint src);`	**UNPKHU4**	Unpacks the two high unsigned 8-bit values into unsigned packed 16-bit values
`uint _unpklu4(uint src);`	**UNPKLU4**	Unpacks the two low unsigned 8-bit values into unsigned packed 16-bit values
`uint _xpnd2(uint src);`	**XPND2**	Bits 1 and 0 of src are replicated to the upper and lower halfwords of the result, respectively.
`uint _xpnd4(uint src);`	**XPND4**	Bits 3 through 0 are replicated to bytes 3 through 0 of the result.

A.5 Optimization Checklist†

Phase	Description
1	Compile and profile native C code • Validates original C code • Determines which loops are most important in terms of MIPS requirements
2	Add constant declarations and loop count information • Reduces potential pointer aliasing problems • Allows loops with indeterminate iteration counts to execute epilogs
3	Optimize C code using intrinsics and other methods • Facilitates use of certain C6x instructions to be used • Optimizes data flow bandwidth
4	Write linear assembly • Allows control in determining exact C6x instructions to be used • Provides flexibility of hand-coded assembly without pipelining, parallelism, or register allocation
5	Add partitioning information to the linear assembly • Can improve partitioning of loops when necessary • Can avoid bottlenecks of certain hardware resources

About the Author

Nasser Kehtarnavaz received his Ph.D. degree in Electrical and Computer Engineering from Rice University in 1987. He joined the Department of Electrical Engineering at Texas A&M University in 1986 as an Assistant Professor where he later became an Associate Professor, and a Professor. He has been a Professor of Electrical Engineering at the University of Texas at Dallas since 2002. Dr. Kehtarnavaz's research areas include signal and image processing, real-time imaging, DSP-based system design, biomedical image analysis, and pattern recognition. He has authored or co-authored three books and numerous journal and conference papers in these areas. Among his many professional activities, he is currently serving as the Editor-in-Chief of Journal of Real-Time Imaging, and Chair of the Dallas Chapter of the IEEE Signal Processing Society. Dr. Kehtarnavaz is a Fellow of SPIE, a Senior Member of IEEE, and a Professional Engineer.

Index

ELSEVIER SCIENCE CD-ROM LICENSE AGREEMENT

TERM

This Agreement will remain in effect until terminated pursuant to the terms of this Agreement. You may terminate this Agreement at any time by removing from Your system and destroying the CD-ROM Product. Unauthorized copying of the CD-ROM Product, including without limitation, the Proprietary Material and documentation, or otherwise failing to comply with the terms and conditions of this Agreement shall result in automatic termination of this license and will make available to Elsevier Science legal remedies. Upon termination of this Agreement, the license granted herein will terminate and You must immediately destroy the CD-ROM Product and accompanying documentation. All provisions relating to proprietary rights shall survive termination of this Agreement.

LIMITED WARRANTY AND LIMITATION OF LIABILITY

NEITHER ELSEVIER SCIENCE NOR ITS LICENSORS REPRESENT OR WARRANT THAT THE INFORMATION CONTAINED IN THE PROPRIETARY MATERIALS IS COMPLETE OR FREE FROM ERROR, AND NEITHER ASSUMES, AND BOTH EXPRESSLY DISCLAIM, ANY LIABILITY TO ANY PERSON FOR ANY LOSS OR DAMAGE CAUSED BY ERRORS OR OMISSIONS IN THE PROPRIETARY MATERIAL, WHETHER SUCH ERRORS OR OMISSIONS RESULT FROM NEGLIGENCE, ACCIDENT, OR ANY OTHER CAUSE. IN ADDITION, NEITHER ELSEVIER SCIENCE NOR ITS LICENSORS MAKE ANY REPRESENTATIONS OR WARRANTIES, EITHER EXPRESS OR IMPLIED, REGARDING THE PERFORMANCE OF YOUR NETWORK OR COMPUTER SYSTEM WHEN USED IN CONJUNCTION WITH THE CD-ROM PRODUCT.

If this CD-ROM Product is defective, Elsevier Science will replace it at no charge if the defective CD-ROM Product is returned to Elsevier Science within sixty (60) days (or the greatest period allowable by applicable law) from the date of shipment.

Elsevier Science warrants that the software embodied in this CD-ROM Product will perform in substantial compliance with the documentation supplied in this CD-ROM Product. If You report significant defect in performance in writing to Elsevier Science, and Elsevier Science is not able to correct same within sixty (60) days after its receipt of Your notification, You may return this CD-ROM Product, including all copies and documentation, to Elsevier Science and Elsevier Science will refund Your money.

YOU UNDERSTAND THAT, EXCEPT FOR THE 60-DAY LIMITED WARRANTY RECITED ABOVE, ELSEVIER SCIENCE, ITS AFFILIATES, LICENSORS, SUPPLIERS AND AGENTS, MAKE NO WARRANTIES, EXPRESSED OR IMPLIED, WITH RESPECT TO THE CD-ROM PRODUCT, INCLUDING, WITHOUT LIMITATION THE PROPRIETARY MATERIAL, AN SPECIFICALLY DISCLAIM ANY WARRANTY OF MERCHANTABILITY OR FITNESS FOR A PARTICULAR PURPOSE.

If the information provided on this CD-ROM contains medical or health sciences information, it is intended for professional use within the medical field. Information about medical treatment or drug dosages is intended strictly for professional use, and because of rapid advances in the medical sciences, independent verification f diagnosis and drug dosages should be made.

IN NO EVENT WILL ELSEVIER SCIENCE, ITS AFFILIATES, LICENSORS, SUPPLIERS OR AGENTS, BE LIABLE TO YOU FOR ANY DAMAGES, INCLUDING, WITHOUT LIMITATION, ANY LOST PROFITS, LOST SAVINGS OR OTHER INCIDENTAL OR CONSEQUENTIAL DAMAGES, ARISING OUT OF YOUR USE OR INABILITY TO USE THE CD-ROM PRODUCT REGARDLESS OF WHETHER SUCH DAMAGES ARE FORESEEABLE OR WHETHER SUCH DAMAGES ARE DEEMED TO RESULT FROM THE FAILURE OR INADEQUACY OF ANY EXCLUSIVE OR OTHER REMEDY.

U.S. GOVERNMENT RESTRICTED RIGHTS

The CD-ROM Product and documentation are provided with restricted rights. Use, duplication or disclosure by the U.S. Government is subject to restrictions as set forth in subparagraphs (a) through (d) of the Commercial Computer Restricted Rights clause at FAR 52.22719 or in subparagraph (c)(1)(ii) of the Rights in Technical Data and Computer Software clause at DFARS 252.2277013, or at 252.2117015, as applicable. Contractor/Manufacturer is Elsevier Science Inc., 655 Avenue of the Americas, New York, NY 10010-5107 USA.

GOVERNING LAW

This Agreement shall be governed by the laws of the State of New York, USA. In any dispute arising out of this Agreement, you and Elsevier Science each consent to the exclusive personal jurisdiction and venue in the state and federal courts within New York County, New York, USA.